Corus UK Limited
Swinden Technology Centre

Library and Information Services
Moorgate, Rotherham   S60 3AR

NOTES TO BORROWERS

1. This publication should be returned or the loan renewed within 21 days, i.e. by the last date shown below.

2. As this Library is primarily a reference tool for the Technology Centre, borrowed publications should be readily accessible on demand.

3. It is a condition of this loan that, in case of damage or loss, this publication shall be replaced at the cost of the borrower

Name	Address	Date Due	Initial Before Returning
Ben Scotham		22.10.10	

D1341163

FUNDAMENTAL ENGINEERING MATHEMATICS:

A Student-Friendly Workbook

"Talking of education, people have now a-days" (said he) "got a strange opinion that every thing should be taught by lectures. Now, I cannot see that lectures can do so much good as reading the books from which the lectures are taken. I know nothing that can be best taught by lectures, except where experiments are to be shewn. You may teach chymestry by lectures — You might teach making of shoes by lectures!"

James Boswell: *Life of Samuel Johnson, 1766*

ABOUT THE AUTHORS

Neil Challis was born in Cambridge, UK. He studied mathematics at the University of Bristol and subsequently worked for some years as a mathematician in the British Gas Engineering Research Station at Killingworth. Since 1977, he has worked in the Mathematics Group, Sheffield Hallam University, UK and is currently head of that group. He obtained a PhD in mathematics from the University of Sheffield in 1988 and has taught mathematics to a wide variety of students, across the spectrum from first year engineers and other non-mathematicians who need access to mathematical ideas, techniques and thinking, to final year single honours mathematics students.

Harry Gretton was born in Leicester, UK. He studied mathematics at the University of Sheffield, obtaining his PhD from there in 1970. He has taught mathematical sciences since then, both at Sheffield University and Sheffield Hallam University, and has been a tutor with the Open University since it was conceived. He has taught many varied students on many varied mathematically-related courses. In recent years he has developed a particular interest in the impact of technology on the way mathematics is taught, practiced and assessed.

FUNDAMENTAL ENGINEERING MATHEMATICS:
A Student-Friendly Workbook

Neil Challis
Head of Mathematics and Statistics
Sheffield Hallam University

Harry Gretton
Principal Lecturer in Mathematics and Statistics
Sheffield Hallam University

Horwood Publishing
Chichester, UK

HORWOOD PUBLISHING LIMITED
International Publishers in Science and Technology
Coll House, Westergate, Chichester,
West Sussex, PO20 3QL England

First published in 2008

British Library Cataloguing in Publication Data
A catalogue record of this book is available from the British Library

ISBN – 10: 1-898563-65-9
ISBN – 13: 978-1-898563-65-5

Printed and bound in the UK by CPI Antony Rowe.

To you, the student of engineering using this book:

This is a **workbook**, not a textbook. We think it can help you if you are on a foundation or first year degree course. You learn mathematical skills by *doing* mathematics - not by reading about what others have done, although others can guide you. So *do the exercises and activities as you come to them!*

Do your mathematics with your tools and technology next to you. These should perhaps include pen and paper, calculator, graphic calculator, spreadsheet, Computer Algebra System (CAS) such as Derive, Mathcad, Maple, Mathematica, or TI-Interactive. This is not however an instruction manual for particular technologies, as these go out of date quickly. Instead, we encourage you to learn to use whatever tools you have to get a rich view of what mathematics is about, and to get your understanding and your answers.

With all this technological power, you have multiple ways of getting answers, so you should never "get the sums wrong". Your technology is very good at repetitive calculations and giving answers, but it cannot think for you.
- Many of the most common mistakes come from simple things such as not entering expressions properly into your technology. Remember BODMAS, and work through the early chapters carefully, even if you think you know this "easy" stuff already. Work on your understanding of the *meaning* of algebraic expressions. Mathematics is a language with meaning, not just a collection of algebraic tricks.
- Never take an answer at face value until you have checked it independently and can convince yourself that you are right and your answer is sensible. To this end, we do not give many answers to exercises, but will suggest ways in which you can check your own.

You will find that you might take an algebraic, or a graphical or a numerical approach. You may also have to describe what is happening. All these skills and ways of thinking are useful to you as an engineer. We call it going for a **SONG** - a mix of **S**ymbolic, **O**ral, **N**umerical and **G**raphic approaches.

Be conscious of how you are learning. Use mind tricks such as Predict-Observe-Explain: predict an answer by one means (perhaps on paper or by drawing a picture), observe what the answer is (perhaps using some technology) and explain to yourself, and to anyone else who will listen, about any discrepancies.

SO … do the examples and check what you do by whatever means you have. Talk to people about what you are doing, both if you are pleased and if you are stuck. Mix old (paper) and new (electronic) technology. Use algebra, use numbers, draw pictures, and learn to interpret what these mean. Look for understanding, not just for learning tricks without knowing why they work. Above all, *ENJOY IT*!

TABLE OF CONTENTS

1

Numbers, Graphics and Algebra

"Can you do addition?" the White Queen asked. "What's one and one and one and one and one and one and one and one and one and one?" "I don't know," said Alice. "I lost count." Lewis Carroll: *Through the Looking Glass*

1.1 NUMBERS, GRAPHICS AND ALGEBRA

The purpose of these first few chapters is to get you to look with fresh eyes at some basic mathematical ideas that you might think you already know all about. You will see how technology can help you to do that and can help to develop a deeper understanding of those ideas. You need this understanding and "feel" for what numbers and pictures can tell you about the world. You will almost certainly see "doing maths" differently from the way you saw it earlier in your life.

This chapter does not give a complete treatment for beginners in its topics. It provides a fresh look at, and reminder of, a mixture of ideas that you will find useful. You will think again about numbers and how they behave and how you can very simply represent them by letters – the idea of algebra. You will revisit simple geometry, as well as the ideas of trigonometry, describing the links between distances and angles. You will also develop ideas about the strong links between numbers, algebra and pictures - the pictures are what bring the numbers to life.

As you work through this chapter, and those which follow, remember that mathematics is a "doing", not a "watching" activity, so make sure you keep on **doing the exercises and activities** as you go.

1.2 WHAT NUMBERS ARE

First you need to remind yourself of the basic terminology of numbers – bearing in mind that in mathematics some words have a precise meaning which is more specialised than in standard English.

Natural numbers are the positive whole numbers 1, 2, 3 ... (0 may be included).

Integers are *all* whole (including negative) numbers … -2, -1, 0, 1, 2, 3 … .

Rational numbers are those which can be written as fractions in the form $\frac{a}{b}$ where *a* and *b* are integers.

Some important and useful numbers cannot be written in rational form – e.g. the special numbers $\pi \approx 3.14159..$, $e \approx 2.71828..$ and $\sqrt{2} \approx 1.41421..$, which you will meet later, if you have not already done so. Such numbers are called **irrational** and occur with infinitely many non-repeating decimal places.

When the rational and irrational numbers are combined, they are called **real numbers** and these are what you will normally think of as decimal numbers.

It is useful to think of numbers graphically being represented by their position on a horizontal straight line, with distance away from a fixed point O governed by their value, and with negative numbers to the left and positive to the right (see Figure 1.1).

Figure 1.1 The number line

1.3 HOW NUMBERS (AND LETTERS) BEHAVE

You will be familiar with the arithmetic operations – add, subtract, multiply, divide, exponentiation (raise to a power), and the use of brackets. You should of course be used to using your calculator (graphic or otherwise) to do such things, but you will find that even with technology, if we do not all agree to use the same rules and language, then you can have serious problems. You must communicate properly, not only with other people, but also with machines.

First let us ask you a question in words – answer it **without** using your calculator: "What is one plus two times three?" **DO IT NOW!**

Some of you may get the answer 7, and others may get 9. Can you explain why two answers are possible? Check it on your calculator. You see that we need to agree the order of operations: if you do the addition first, you get 9; doing the multiplication first gives 7.

Thus, in this section, we give you the chance to remind yourself of the rules. We must all agree to use BODMAS, which some remember as Please Excuse My Dear Aunt Sally, to tell us the order in which to carry out operations:

    BODMAS (**B**rackets/**O**rder/**D**ivision/**M**ultiplication/**A**ddition/**S**ubtraction)

    PEMDAS (**P**lease **E**xcuse **M**y **D**ear **A**unt **S**ally -

    Parentheses/**E**xponents/**M**ultiplication/**D**ivision/**A**ddition/**S**ubtraction) .

This means that brackets can override any other order of operation by forcing calculation of quantities in brackets first; then powers are done; then multiply and divide at the same level of priority working from left to right, and, finally, add and subtract working from left to right.

DO THIS NOW! 1.1
Investigate **the order in which your calculator does numerical operations**, by looking at the following expressions (note we use * here for "multiply" and / for "divide", just as many technologies do). Recall two negative quantities multiplied together make a positive. Set your calculator accuracy (number of decimal places) to a sensible level before you start. First, predict without technology which of the expressions in column **A** is equal to the corresponding expression in column **B**. Look particularly at the effect of brackets. Then use your calculator to check whether you are right, explaining any discrepancies.

A	B
1 + 2	2 + 1
1 − 2	2 − 1
(1 + 2) + 3	1 + (2 + 3)
(1 − 2) − 3	1 − (2 − 3)
3 * 4	4 * 3
3 /4	4 /3
7 + 3*2	(7 + 3)*2
-3 + 4	-(3 + 4)
6/3 + 4	6/(3 + 4)
(3/4)/5	3/(4/5)

Notice if you have a graphic calculator, it probably has two different "-" keys. One will be the unary, or leading minus, often made with three pixels, as in (-1). The other is the take-away minus sign – which often has five pixels. See Figure 1.2 to see the unary minus in action. Sort this out on your personal technology **NOW**.

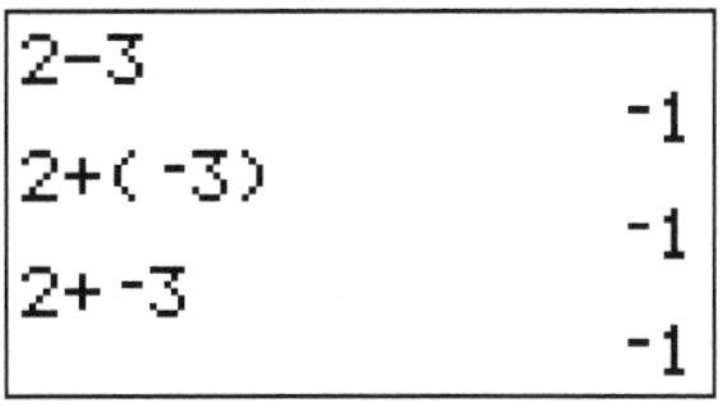

```
2-3
              -1
2+( -3)
              -1
2+ -3
              -1
```

Figure 1.2 The unary minus in action

DO THIS NOW! 1.2
Predict by hand the value your calculator will give for the following expression, and then check by doing it on your calculator, explaining, step by step, what your calculator has done to evaluate this and what rule, as well as BODMAS, it is using:

$$2*3*4/5/6/7 + 8 - 6 = ?$$

1.4 FRACTIONS, DECIMALS AND SCIENTIFIC NOTATION

Although you probably studied fractions years ago, we find that many people have forgotten the rules, or worse still, think they remember them but do not. It is true that you can do fractional number calculations on your calculator, and get your answer as either a fraction or a decimal, and that you can deal with algebraic fractions using a **CAS** (Computer Algebra System) such as Derive. However, it is part of getting a feel for what is happening, to be able to do some simple manipulations with algebraic expressions involving fractions, so here is an opportunity to do just that.

First, here is a quick reminder of the rules:

To **add** or **subtract** two fractions, put them over a common denominator (**bottom**) so that you are adding or subtracting like with like.

Example 1.4.1 $\quad \dfrac{1}{2}+\dfrac{2}{3}=\dfrac{3}{6}+\dfrac{4}{6}=\dfrac{7}{6}$ $\qquad$ and $\qquad$ $\dfrac{2}{3}-\dfrac{1}{2}=\dfrac{4}{6}-\dfrac{3}{6}=\dfrac{1}{6}$

To **multiply** two fractions, multiply the numerators (**tops**) and multiply the denominators (bottoms). To **divide** two fractions, just turn the second one upside down and multiply!

Example 1.4.2 $\quad \dfrac{1}{2}\times\dfrac{2}{3}=\dfrac{1\times2}{2\times3}=\dfrac{2}{6}=\dfrac{1}{3}$ $\qquad$ and $\qquad$ $\dfrac{1}{2}\div\dfrac{2}{3}=\dfrac{1}{2}\times\dfrac{3}{2}=\dfrac{1\times3}{2\times2}=\dfrac{3}{4}$

Note you can simplify a fraction by multiplying or dividing top and bottom by the same number:

Example 1.4.3 $\quad \dfrac{3}{9}=\dfrac{1}{3}$, dividing top and bottom by 3. This is commonly called **cancelling**. However remember that you **cannot** simply *add* or *subtract* numbers on top and bottom – a quick example should convince you of that:

Example 1.4.4 $\quad \dfrac{4+1}{9+1}=\dfrac{5}{10}=\dfrac{1}{2}$ is **not** the same as $\dfrac{4}{9}$. That is to say you cannot cancel the +1 terms if they are added (or subtracted).

DO THIS NOW! 1.3

Evaluate the following expressions by hand, giving your answer as a single fraction. In each case, check with your calculator, considering carefully how you must type things in to take account of My Aunt Sally and so on, and finding out how to make your calculator give you the answer as both a decimal and a fraction.

(a) $\dfrac{2}{5}+\dfrac{3}{7}$ $\qquad\qquad\qquad\qquad$ (b) $\dfrac{2}{5}-\dfrac{3}{7}$

(c) $\dfrac{2}{5}\times\dfrac{3}{7}$ $\qquad\qquad\qquad\qquad$ (d) $\dfrac{2}{5}\div\dfrac{3}{7}$

Make up extra problems of your own if you need more practice, checking your own answers on your calculator.

To close this section, here are a few miscellaneous points to note.

*   Fractions are converted into decimals by dividing the numerator by the denominator.

*   A fraction is converted into percentage form by multiplying its decimal form by 100 and writing % after the result.

*   A ratio is one of the ways of comparing one thing with another. This can be done for instance by comparing part with part. For example two quantities in a ratio 15:20 make up fractional amounts $\frac{15}{35}$ and $\frac{20}{35}$ of the whole.

*   To round to a given number of **decimal places**, go to the decimal place specified. If the next digit to the right is 5 or more, then round the number in the specified final decimal place up; if it is less than 5 then round down. To round to a given number of **significant places**, start counting significant figures from the first non-zero digit on the left and follow the rules for rounding.

*   In **scientific notation**, positive numbers are expressed in the form $a \times 10^n$ where a is between 1 and 10 and n is an integer. Note the use of E in most technologies to represent the power of 10 – see Figure 1.3 for a graphic calculator representation of the number (in scientific notation) $1.862996833 \times 10^{-4}$.

```
23/123457
    1.862996833E-4
```

Figure 1.3 A typical calculator representing a number in scientific notation

*   For any numerical problem always give the answer in the appropriate form and accuracy. For instance if your data are recorded to one decimal place then it is inappropriate to give an answer with more places.

DO THIS NOW! 1.4

Express the following fractions $\frac{1}{7}$, $\frac{2}{3}$, $\frac{123}{43}$, $\frac{23}{123457}$:

(i) as percentages,   (ii) as decimals rounded to 6 decimal places,

(iii) to three significant figures,   (iv) in scientific notation.

1.5 POWERS OR INDICES

You need to remind yourself about how powers or indices behave. Recall that if n is a natural number, a^n means do $a \times a \times a \dots \times a$ where a occurs n times. a is called the base, and n is called the index or power. Simple rules for combining powers emerge from this:

$$a^m \times a^n = a^{m+n}$$ If you *multiply* the numbers, you *add* the powers.

$$\frac{a^m}{a^n} = a^{m-n}$$ If you *divide* the numbers, you *subtract* the powers.

$$(a^m)^n = a^{m\times n}$$ *Raise to one power after another: multiply* the powers.

These simple rules lead to natural meanings for powers other than natural numbers.

$$a^0 = 1$$ This is because $\frac{a^m}{a^m} = 1$ (obviously?), but also

$$a^{-n} = \frac{1}{a^n}$$ This is because $\frac{1}{a^n} = \frac{a^0}{a^n} = a^{0-n} = a^{-n}$.

$$a^{\frac{1}{2}} = \sqrt{a}$$ This is because $a^{\frac{1}{2}} \times a^{\frac{1}{2}} = a^{\frac{1}{2}+\frac{1}{2}} = a^1$.

Thus $a^{\frac{1}{2}}$ behaves like a square root.

Similarly $a^{\frac{1}{n}} = \sqrt[n]{a}$ is the nth root.

DO THESE NOW! 1.5
For each expression which follows, work on paper using the appropriate law to express it as a single power of the base, with no fraction. You do not need answers because you can check using your calculator whether you are right.

(i) $2^3 \times 2^4$ (ii) $\dfrac{1}{9^2}$ (iii) $1.2^{3.4} \times 1.2^{5.7}$ (iv) $\dfrac{1}{5.67^{-2.3}}$

(v) $\dfrac{3^9}{3^4}$ (vi) $(5^4)^3$ (vii) $\dfrac{4.5^{3.4}}{4.5^{2.2}}$ (viii) $(6.7^{3.5})^{7.8}$.

1.6 ANGLE AND LENGTH - GEOMETRY AND TRIGONOMETRY
1.6.1 Angle measurement

Having spent some time looking again at numbers, you should now remind yourself of some of the basics about mathematical diagrams - in this case geometry and trigonometry. First, look at angles. There are two common unit systems for measuring these, the first being **degrees**. The Babylonians divided the circumference of a circle into 12 sections, and divided each of these sections into 30 equal parts – thus creating 360 divisions around the circle, each corresponding to one degree (1°).

The other angle unit is the **radian**. The angle formed at the centre of a circle by an arc equal in length to the radius is defined to be one radian (1 rad $\equiv 1^c$). See Figure 1.4. Thus there are 2π radians in one full revolution and therefore $360^0 \equiv 2\pi^c$, so that is 360 degrees is the same as 2π radians.

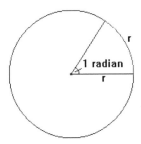

Figure 1.4 The definition of a radian

1.6.2 Triangles and connections between length and angle

A **triangle** is a polygon with three sides. The sum of its internal angles is 180^0 or π radians. If all three sides are the same length then it is called an **equilateral** triangle and all angles are 60^0 or $\dfrac{\pi}{3}^c$. If two sides are of equal length then it is an **isosceles** triangle and the two angles opposite the equal length sides are equal. The area of a triangle is $\dfrac{1}{2}ah$ where a is the length of the base and h is the height. This can also be written as $\dfrac{1}{2}ab\sin\theta$ where a and b are two side lengths and θ the angle between those sides.

1.6.3 Right-angled triangles and basic trigonometry

A right-angled triangle is one such as ABC, as shown in Figure 1.5, where in this case C is the right angle and sides of lengths a, b, and c are opposite the angles A, B and C. The side opposite the right angle is known as the **hypotenuse**.

One special relationship here is given by **Pythagoras' Theorem**. This tells us that for a triangle such as ABC, the lengths are related by the formula $a^2 + b^2 = c^2$.

You should also remind yourself of the trigonometric ratios. These can be defined initially in a right-angled triangle such as that in Figure 1.5, as below.

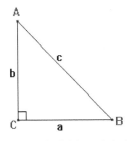

Figure 1.5 A standard right-angled triangle

Sine	Cosine	Tangent
$\sin(A) = \dfrac{a}{c} = \dfrac{opposite}{hypotenuse}$	$\cos(A) = \dfrac{b}{c} = \dfrac{adjacent}{hypotenuse}$	$\tan(A) = \dfrac{a}{b} = \dfrac{opposite}{adjacent}$

There are many relationships between these; for instance start from Pythagoras' Theorem $a^2 + b^2 = c^2$ and divide both sides by c^2. This gives $\dfrac{a^2}{c^2} + \dfrac{b^2}{c^2} = 1$ or in other words $\sin^2 A + \cos^2 A = 1$.

DO THIS NOW! 1.6

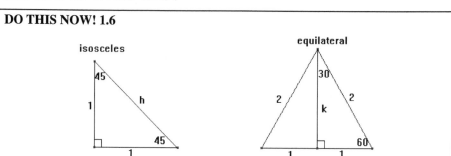

Figure 1.6 Trigonometric ratios on special triangles

For the triangles in Figure 1.6, use Pythagoras' theorem to work out the values of lengths h and k giving your answer in terms of the square roots of 2 and 3.

Using these triangles and the definitions of sine, cosine and tangent, complete the following table giving values in terms of the square roots as above. Check your answers with your calculator, making sure you are in the right mode!

Angle A in degrees	0	30	45	60	90
Angle A in radians	0	$\dfrac{\pi}{6}$	$\dfrac{\pi}{4}$	$\dfrac{\pi}{3}$	π
$\sin(A)$					
$\cos(A)$					
$\tan(A)$					

END OF CHAPTER 1 - CALCULATOR ACTIVITIES – DO THESE NOW!

Activity 1.1 Warm-up basic calculator exercises

If you are asked to find the value of $\frac{5}{12} + \frac{2}{3} + \frac{3}{4}$, you might do one of the two things shown on the calculator screen in Figure 1.7:

```
5/12+2/3+3/4
              1.83333333
5/12+2/3+3/4▶Fra
c
              11/6
```

Figure 1.7 Fractions or decimals?

The first answer, presented in decimal form, shows you are competent, but the second answer shows you understand that it is more appropriate to give your answer as a fraction, since that is how the question is posed. Always think about the sensible form in which to present your answers – fraction or decimal, accuracy, etc.

Now evaluate, in appropriate form, the answers to (a) – (h) below, to help you get comfortable with your calculator or CAS (Computer Algebra System), and to check that you can handle these basic operations with some confidence. Think about how to verify your own answers, and do so before checking our answers.

(a) $\dfrac{9}{24} \times 6 + \dfrac{11}{12} \times 4$

(b) Subtract $5\frac{3}{8} + 2\frac{5}{7}$ from $8\frac{5}{8} - 2\frac{7}{9}$

(c) $\dfrac{3\frac{3}{5} \times \frac{5}{9} - \frac{7}{19}}{\frac{2}{7} \times 4\frac{3}{8} - \frac{1}{12} \times 5\frac{3}{4}}$

(d) $\dfrac{0.056}{0.15} \times \dfrac{0.035}{0.0023} \div \dfrac{0.14}{4.5}$

(e) $\dfrac{0.203 \times 0.005 \times 1.7}{0.006 \times 0.0029}$

(f) $\dfrac{1.23 + 456.7 + 8.9101}{51.34 - 4.04}$

(g) $\sqrt{18 \times 19 \times 567 \div 29}$

(h) $\sqrt{\dfrac{25}{28} + 1\frac{3}{8} + 2\frac{1}{4} - 3.57}$

Do you agree with our answers just below here? Are you convinced? What form have you used? Why?

(a) $71\frac{1}{12}$ (b) $-565\frac{5}{252}$ (c) $1488\frac{8}{703}$ (d) 182.61 (e) 99.17 (f) 9.87 (g) 81.77 (h) 0.97

Activity 1.2 More warming up with your calculator

The graphic calculator screens in Figure 1.8 show the importance of being able to enter your expressions properly into your calculator – just two of several ways to enter powers. **Learn NOW how *yours* works by tackling these activities below and always remember BODMAS.**

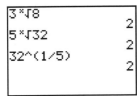

Figure 1.8 "Talking" properly with your calculator

1.2.1 Evaluate (a) to (j), using your calculator and checking appropriately:

(a) 7^3 (b) 3^7 (c) 5^{-2} (d) 2^{-5} (e) $(-2)^6$

(f) -2^6 (g) $(1.234)^8$ (h) 2.345^{-5} (i) $(-6.78)^{-4}$ (j) -6.78^{-4}

1.2.2 Evaluate (a) – (d), using your calculator and checking appropriately:

(a) $2^3 \times 2^5 \times 2^7$ (b) $3^3 \times 3^{-4}$ (c) $4^6 \times 4^{-7} \div 4^{-5}$ (d) $\dfrac{5^5 \times 5^{-3}}{5^4 \times 5^{-2} \times 5^{-7}}$

1.2.3 Use the laws of indices to express (a) – (h) as single powers of 2:

(a) $2^5 \times 2^7$ (b) $2^{10} \div 2^7$ (c) $2^5 \div 2^4$ (d) $2^3 \div 2$

(e) $2^4 \div 2^4$ (f) $(2^3)^4$ (g) $(2^4)^3$ (h) $(2^4)^4$

1.2.4 Use the laws of indices to express (a) – (h) as single powers:

(a) $y^5 \times y^8$ (b) $x^{10} \div x^7$ (c) $c^7 \div c^4$ (d) $x^3 \div x$

(e) $y^4 \div y^6$ (f) $(a^3)^2$ (g) $(a^2)^3$ (h) $(d^4)^5$

Activity 1.3 Using your calculator properly – food for thought

Whether you are getting solutions by hand or machine you should always be sceptical and question your answers until you convince yourself they are right. Many times your pencil and paper or machine will give you an answer which may be incorrect. Get in the habit of verifying what you are doing in at least two ways – for instance, hand calculation and at least one technology. From this comes conviction if they agree, or understanding if you have to find out why there was a difference. Read on ...

As a warning about blindly trusting your technology, look at the graphic calculator screens in Figure 1.9:

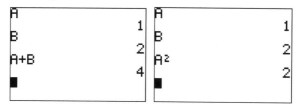

Figure 1.9 Never let you calculator do the thinking for you!

What has gone wrong? Surely $1 + 2 = 3$ and $1^2 = 1$? **THINK NOW** about what may have happened *before* looking at Figure 1.10 for the explanation.

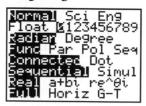

Figure 1.10 Know your technology!

Figure 1.10 shows the mode screen for the calculator (how it is set up) and it reveals through the **Float** line that the machine is set to display numbers to **0** decimal places! In fact A is 1.4 and B is 2.4, and you are seeing the effect of rounding – NOW work out the precise effect this is having and never forget it!

2

Linking Algebra and Graphics 1

"... patterns, like the painter's or the poet's, must be beautiful; the ideas, like the colours or the words, must fit together in a harmonious way". G. H. Hardy

2.1 ALGEBRA AND PICTURES

You should have met the idea of algebra before – the practice of using letters to represent constant or variable quantities, which proves so useful in describing engineering systems. We assume that you have. However many people find the rules for handling algebraic symbols difficult to manage. One reason may be that you try to remember the rule rather than to understand the principle behind the rule. In this chapter we try to give you space to revisit and expand some of your knowledge of algebra. By reading on and doing the exercises, we hope you will gain insight into algebraic processes; give meaning to algebra by linking it to pictures; and investigate how technology can help you to become confident with your algebra.

You will handle symbols, constants, variables, brackets, powers, and algebraic fractions. You will review the difference between an expression and an equation. You will look at solving an equation, in a variety of ways, including through algebraic manipulation, by pictures, by numbers, and using technology. When dealing with pictures, we will restrict ourselves to those involving straight lines only in this chapter. Finally, you will use these ideas with some engineering formulae.

Remember to keep on **doing the exercises and activities** as you reach them.

2.2 NUMBERS, LETTERS AND BRACKETS

Here are a few exercises to get you into simple manipulations, involving numbers, letters and brackets. It is not a comprehensive algebra course but should get you thinking about what you have met and becoming more confident with it.

2.2.1 Multiplying out brackets

First recall how to **multiply out brackets**. Remember the number outside the bracket multiplies every added or subtracted term inside the bracket; remember also that if you multiply two negative terms together, you get a positive. Be careful when you attempt supposedly simple calculations – these are where it is easiest to make a mistake! If ever you are in any doubt when doing simple algebra, think whether what you are doing would work with numbers.

Example 2.2.1

(a) First with numbers only:

Simplify $5(6 + 2) - 3(4 - 6) = 5 \times 6 + 5 \times 2 - 3 \times 4 - 3 \times (- 6) = 30 + 10 - 12 + 18 = 46$.

(b) Now substitute with a letter, as follows:

Simplify $5(t + 2) - 3(4 - t) = 5t + 5 \times 2 - 3 \times 4 - 3 \times (- t) = 5t + 10 - 12 + 3t = 8t - 2$.

Next, expand these ideas to multiply two brackets together.

Example 2.2.2

(a)   What is $(3 + 6) \times (1 + 2)$? Adding the numbers in the brackets first, the answer is of course $(9) \times (3) = 27$. Can we get the same answer by multiplying out the brackets? Yes, by forming the sum of all possible cross-multiplications of terms inside the brackets. One way of making sure this works is to use **FOIL**:

First Outer Inner Last

(and don't forget My Dear Aunt Sally, still here after Chapter 1.)

Thus $(3 + 6) \times (1 + 2) = 3 \times 1 + 3 \times 2 + 6 \times 1 + 6 \times 2 = 3 + 6 + 6 + 12 = 27$ again.

F   O   I   L

Check this again by typing the expression $(3 + 6)(1 + 2)$ into your calculator. Does your machine need a multiplication sign (perhaps *) between the brackets, or will it understand that when you write two brackets next to each other, you mean "multiply"? **FIND OUT NOW!**

(b) Now here is a similar example with algebra.

$(a+b) \times (c+d) = (a+b)(c+d) = ac + ad + bc + bd$.

F   O   I   L

2.2.2 Simple factorisation

Now recall how to **put brackets into expressions** – that is, to **factorise**. People generally find this harder than multiplying out brackets.

Example 2.2.3

You can factorise simple expressions simply by looking for common factors through all terms. For example:

(a) $5x + 10y = 5(x + 2y)$, taking the common factor of 5 outside the bracket. Check this by multiplying out the bracket again to see if it works in reverse.

(b) $3ab^2c + 9ab = 3ab(bc + 3)$ using the common factor of $3ab$.

Once more you can check by multiplying out the bracket. Another way to check is to use a Computer Algebra System (CAS) such as Derive, Mathcad, Maple or perhaps your graphic calculator. Explore what tools you have available **NOW**.

DO THIS NOW! 2.1

1. Multiply out these brackets. In each case check your answer (a) by substituting numbers for the letters and seeing if the answer works, and (b) by using your local CAS package – **FIND OUT NOW** how to enter an algebraic expression and expand it (that is, multiply it out).

(a) $5(t+2) - 3(4-t)$ (b) $(a+1)(b+2)$ (c) $(a+b)(b+c)$

(d) $(2a+b)(a-b)$ (e) $(2x-y)(x-3y)$.

2. Put these expressions into factorised form as far as possible. In each case check your answer by (a) multiplying the brackets back out and seeing if you get back to what you started with, and (b) by using your local CAS package – **FIND OUT NOW** how to enter an algebraic expression and factorise it:

(a) $3a+6b$ (b) $5a^2 - 25ab$ (c) $12ab+15ab^2$

(d) $24xyz - 16x^2 y + 20xy^2$ (e) $ac+ad+bc+bd$.

2.3 "SPEAKING" ALGEBRA

One very useful skill is to be able to take problems expressed in words, and turn them into algebraic problems. For instance:

Example 2.3.1

 (a) From the sum of $2a^2 + ab - b$ and $2ab+3b$, subtract $a^2 - ab+b$ and simplify your answer as far as you can:

$$(2a^2 + ab - b)+(2ab+3b) - (a^2 - ab+b) = a^2 + 4ab+b .$$

 (b) Multiply $a^2 + ab+b^2$ by $a-b$ and simplify your answer as far as you can:

This gives the expression below, which we multiply out by a logical development of **FOIL**:

$$(a^2 + ab+b^2)(a-b) = a^3 - a^2 b+a^2 b - ab^2 +b^2 a - b^3 = a^3 - b^3 .$$

DO THIS NOW! 2.2

Do the following algebraic sums. In each case express the answer as a single algebraic expression, in the simplest form you can find.

1. From the sum of $x^2 - 3x+7$ and $2x^2 + 4x$, subtract $3x^2 + 4$.

2. From the sum of $a+3b-2c$, $3a-b-c$ and $a+6c$, subtract the sum of $a-3b$ and $2b-c$.

3. Multiply $x^2 - 3x+5$ by $2x^2 +6x+2$.

4. Multiply $a^2 - ab+2b^2$ by $2a+b$.

2.4 ALGEBRAIC FRACTIONS

The rules for *algebraic* fractions are the same as for *numerical* fractions.

You can multiply or divide top and bottom of a fraction by the same (non-zero) quantity.

Example 2.4.1 $\dfrac{3a}{5a} = \dfrac{3}{5}$, dividing top and bottom by a (where $a \neq 0$). Remember this is called *cancelling*. Although it is less obvious with letters than numbers, you cannot cancel added or subtracted quantities – see this next example:

Example 2.4.2 $\dfrac{a+1}{b+1} \neq \dfrac{a}{b}$ Verify by putting in numbers – say $a = 2$ and $b = 3$.

To **add** or **subtract** two fractions, put them over a common denominator (bottom) so that you are adding or subtracting like with like – that is, once you have created the common denominator, just add or subtract the tops.

Example 2.4.3 $\dfrac{a}{b} + \dfrac{c}{d} = \dfrac{ad}{bd} + \dfrac{cb}{db} = \dfrac{(ad+cb)}{bd}$ and $\dfrac{a}{b} - \dfrac{c}{d} = \dfrac{ad}{bd} - \dfrac{cb}{db} = \dfrac{(ad-cb)}{bd}$

You must realize that as one fraction has b on the bottom and the other one d, then the common denominator will have to be bd.

To **multiply** two fractions, multiply the numerators and multiply the denominators. To **divide** two fractions, just turn the second one upside down and multiply!

Example 2.4.4 $\dfrac{a}{b} \times \dfrac{c}{d} = \dfrac{a \times c}{b \times d} = \dfrac{ac}{bd}$ and $\dfrac{a}{b} \div \dfrac{c}{d} = \dfrac{a}{b} \times \dfrac{d}{c} = \dfrac{ad}{bc}$

The problem with algebraic fractions is that there are rules for handling them, but to the uncritical eye, correct and incorrect rules can look equally plausible. Be careful, check your answers by more than one means and don't use a rule with letters unless you would use it with numbers and believe it. Also, get used to using brackets on the top and bottom of a fraction. This will often stop you following illogical processes and will also help you to put expressions into your machines.

DO THIS NOW! 2.3

1. In the light of your experience above, express the following as single **algebraic fractions**:

$\dfrac{1}{a} + \dfrac{1}{b} =$ $\qquad\qquad\qquad$ $\dfrac{1}{a} - \dfrac{1}{b} =$

$\dfrac{1}{a} \times \dfrac{1}{b} =$ $\qquad\qquad\qquad$ $\dfrac{1}{a} \div \dfrac{1}{b} =$

2. Simplify the following expressions as far as you can, checking your working with substituted numerical values:

$\dfrac{(2a^2 - 2ab)}{(6ab - 6b^2)} =$ $\qquad\qquad\qquad$ $\dfrac{(x-2)}{(4x-8)} =$

2.5 SOLVING SIMPLE EQUATIONS

Solving equations is a basic mathematical activity with which you need to get to grips. We shall start with so-called linear equations, saying more about why they are called that in a short while. However heavily these are disguised, they involve, at worst, terms like ax (a constant times x), and terms which are just constant. You may have learnt various solution methods in the past, but if you are not confident with these, we suggest you study this area using the ideas below.

Example 2.5.1 Solve, by hand, the equation

$$\frac{2(x+9)}{3} = 15 \ .$$

The rule is to think of the equals sign as a balancing point and so, to keep the equation in balance, whatever you do to the left hand side (LHS) of the equation you must do **exactly** the same to the right hand side (RHS). In general you should:

Get rid of fractions first, by multiplying both sides by something.

Get rid of brackets next by multiplying them out.

Get everything involving the unknown x onto the LHS.

Get everything else onto the RHS.

Leave just x on the LHS by dividing both sides by something.

So

The original equation   is $\dfrac{2(x+9)}{3} = 15$.

Remove fractions by multiplying both sides by 3 $2(x+9) = 45$.

Remove brackets $2x+18 = 45$.

Get x terms on LHS and all else on RHS by subtracting 18 from both sides
$$2x = 45-18 = 27 \ .$$

Divide both sides by 2, to make coefficient of x equal 1 $x = \dfrac{27}{2} = 13.5$.

To check, substitute the answer back into the original equation and see if it really does make the LHS equal to the RHS.

Substitute $x = 13.5$ into the LHS of $\dfrac{2(x+9)}{3} = 15$ to get $\dfrac{2(13.5+9)}{3} = \dfrac{2 \times 22.5}{3} = \dfrac{45}{3}$

which is indeed 15. So the solution works.

Figure 2.1 shows this process can be simply done step by step by a machine to confirm results.

$\blacksquare \dfrac{2\cdot(x+9)}{3}=15$	$\dfrac{2\cdot(x+9)}{3}=15$
$\blacksquare \left(\dfrac{2\cdot(x+9)}{3}=15\right)\cdot 3$	$2\cdot(x+9)=45$
$\blacksquare \text{expand}(2\cdot(x+9)=45)$	$2\cdot x+18=45$
$\blacksquare (2\cdot x+18=45)-18$	$2\cdot x=27$
$\blacksquare \dfrac{2\cdot x=27}{2}$	$x=27/2$

Figure 2.1 Use your technology to check your results.

Many calculators and CAS also have a Solver feature which will solve such equations directly, without having to go through the intermediate steps – but of course you must know how to put your equation in correctly and accurately!

DO THIS NOW 2.4!

Does your calculator have a Solver function? Does your CAS? If so, find out how to use it NOW and check the answer to Example 2.5.1. Does the technology agree with your manual solution? If not, which one is wrong? How can you tell?

For each of the following, first solve by hand and then check your results twice, both on your calculator or CAS, and also by substituting back in the original equation.

(a) $x+8=14$ (b) $x-5=21$

(c) $5x+4=4x+7$ (d) $7x-6=x+12$

(e) $\dfrac{3}{8}(x-2)=2x+4$ (f) $4(y-3)=\dfrac{y}{2}+9$

(g) $2y+1=\dfrac{y}{3}+6$ (h) $d+7=4\dfrac{(1-d)}{3}$.

2.6 CONNECTING STRAIGHT LINES AND LINEAR EXPRESSIONS

Your aim in working through this section should be to remind yourself (or to find out) about the mathematics of straight lines and how the graphs of straight lines are related to linear algebraic equations.

You will need to plot quite a few graphs quickly as you work through this section and there are various ways in which you could do this. We recommend that you use a graphic calculator, but you could use a spreadsheet such as Excel, you could use your CAS, or you could plot them by hand. When you are planning your graph plotting, by hand or by technology, think carefully about what is a sensible scale upon which to plot.

Here is the basic algebraic fact in the form you have probably met it before.

The equation of a straight line can be written in the form $y=mx+c$

m **is the slope (** $\dfrac{vertical}{horizontal}$ **) and** c **is the** y**-intercept (value where line cuts** y**-axis)**

This section consists mainly of activities – do them to recall what these facts mean.

DO THIS NOW! 2.5

Preferably using your technology, plot on the same axes the graphs of

$$y = x \qquad\qquad y = 2x \qquad\qquad y = 3x \ .$$

Predict what the graph of $y = 4x$ will look like and plot it to see if you are right. What do you think $y = 100x$ will look like? Check it out.

Now starting again, plot on the same axes the graphs of

$$y = x \qquad\quad y = -x \qquad\quad y = -2x \qquad\quad y = -3x \ .$$

Predict what the graph of $y = -4x$ will look like and plot it to see if you are right.

Now taking $c = 0$, complete the following sentences by either deleting as appropriate, or completing in your own words:

• If m is positive, then the graph of $y = mx$ slopes ***up/down*** to the right.

• If the graph of $y = mx$ slopes down to the right, then m must be ***positive/negative***.

• If m starts at 0 and increases in a positive direction, the graph of

  $y = mx$ ....

• If m starts at 0 and becomes more negative, the graph of

  $y = mx$ ....

• All lines of the form $y = mx$ pass through the point ....

DO THIS NOW! 2.6

Plot on the same axes the graphs of

$$y = x \qquad\qquad y = x+1 \qquad\qquad y = x+2 \ .$$

Predict what the graph of $y = x+3$ will look like and plot it to see if you are right. What do you think $y = x-1$ will look like? And $y = x-2$? Check them out.

Complete the following sentences by either deleting as appropriate, or adding your own words:

• If c is positive, then $y = mx+c$ cuts the y-axis ***above/below*** the x-axis.

• If the graph of a straight line $y = mx+c$ intersects $x = 0$ below the x-axis then c must be ***positive/negative***.

• If c starts at 0 and increases in a positive direction, the graph of $y = mx+c$

  moves ....

• If c starts at 0 and increases in a negative direction, the graph of $y = mx+c$

  moves ....

• If m is kept constant and c varies, then a family of lines is produced which are

  all ....

Straight lines do not always appear immediately in the standard form $y = mx + c$.

DO THIS NOW! 2.7

Plot the following on the same axes:

$y = x$ $y = x+1$ $y = 2(x+1)$ $y = -3(5-2x)$ $x+y = 1$.

Are they all straight lines? What is it algebraically that makes them so? Try to transform each of them so that they are in the standard form $y = mx + c$.

You should find that this is the common factor that makes them all straight lines. Check your working by verifying that your calculated values for m and c in each case agree with what you observe from the graphs.

2.7 SOLVING LINEAR EQUATIONS GRAPHICALLY

You have seen that a straight line has a standard equation $y = mx + c$ and a linear expression involves, at worst, terms like ax (a constant times x), and terms which are just constant. Since these say the same thing, you might suspect that there is a relationship between the solution of a linear equation and the graph of a straight line, and you would indeed be right. The word "linear" is in fact derived from the word "line".

Example 2.7.1 Do these two things:

* Solve the equation $3x + 21 = 0$.

* Plot the graph of $y = 3x + 21$ and observe where it crosses the x-axis.

You should find the solution to the equation is $x = -7$.

If you plot the graph you should observe that it crosses the x-axis where $x = -7$.

Why are these two values the same? Put simply, the place where the line crosses the x–axis is the place where $y = 0$ and setting $y = 0$ gives exactly the equation we are solving!

DO THIS NOW! 2.8

Check this out by plotting/solving these matching equations and lines (satisfy yourself that they do match first!).

(a) Plot $y = -5x + 16$ and solve $-5x + 16 = 0$.

(b) Plot $y = x - 26$ and solve $x - 5 = 21$.

(c) Plot $y = x - 3$ and solve $5x + 4 = 4x + 7$.

Can you see another way to solve these last two graphically? (Hint. For example with (b), you could see where the straight line $y = x - 5$ produces a y value of 21.)

2.8 TRANSPOSING FORMULAE

This section is here because the algebraic "magic" art of transposing formulae, sometimes called changing the subject of a formula, follows the same rules as equation solving in section 2.5. Sometimes more letters are involved and the expressions are often not as simple. The method is a useful one, even though one of its uses – finding the value of one variable in a formula when you know all the others –is built into much mathematical technology including a CAS and spreadsheet.

Example 2.8.1 The formula for calculating the period T of a simple pendulum of

length L, is:
$$T = 2\pi\sqrt{\frac{L}{g}} \ .$$

where g is the acceleration due to gravity. T is the subject of the formula here.

If we want to change this to make L the subject, then you would just have to follow the same sorts of rules quoted in section 2.5 for solving linear equations.

Divide both sides by 2π
$$\frac{T}{2\pi} = \sqrt{\frac{L}{g}} \ .$$

Square both sides to remove the square root
$$\left(\frac{T}{2\pi}\right)^2 = \frac{L}{g} \ .$$

Multiply both sides by g
$$g\left(\frac{T}{2\pi}\right)^2 = L \ .$$

This approach can work for many but not all formulae. Here is another example.

Example 2.8.2   The elevation e of the outer rail of a railway line is given by $e = \dfrac{4WV^2}{5R}$, where W is the distance between the rails, V is the average speed and R is the radius of the curve. Find V in terms of the other symbols.

A purely symbolic process is necessary here and although we could confirm our answer with a few numbers, this does not guarantee a correct solution. A CAS package could be used to confirm our solution. Thus:

Example 2.8.2 Method 1: Pencil and Paper

Following steps as before, the equation $e = \dfrac{4WV^2}{5R}$ can be rearranged to give

$$\frac{5Re}{4W} = V^2 \Rightarrow V = \pm\sqrt{\frac{5Re}{4W}} \ .$$

Example 2.8.2 Method 2 A CAS system

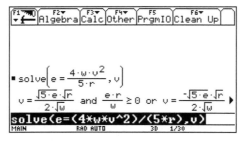

Figure 2.2 Changing the subject of a formula using a TI-89 hand-held, Derive-based CAS

DO THIS NOW! 2.9

In each of the following examples, find your answer using pencil and paper and check using technology.

(a) 1 kg is approximately equal to 2.2 lb (British pound weight). Write this as a formula:

$P = 2.2K$ where K is the mass in kg and P is the mass in lb. By transposing the formula as necessary, use this relationship to convert:

(i)   5 kg to lb

(ii)   6lb to kg.

(b) Using the equation $E = mv^2$, find by hand, transposing the formula as necessary:

(i)   the value of E if $m = 20$ and $v = 7$

(ii)   the value of v if $m = 20$ and $E = 640$.

(c) If $U = V(1 - \dfrac{c}{D\sqrt{N}})$, find N in terms of U, V, D, c.

(d) From the formula $\sqrt{R^2 + (2\pi f L)^2} = \dfrac{V}{I}$, find L in terms of R, f, V and I.

(e) A formula for the safety thickness of a beam is $kt = \sqrt[3]{\dfrac{l^2 W}{b}}$, where t is the thickness, l the length, b the breadth, W the load and k a constant. Obtain the formula for l in terms of the other symbols.

2.9 STRAIGHT LINES IN ENGINEERING

You can describe the behaviour of many engineering systems, either exactly or approximately, by straight lines. Such systems are called LINEAR. Linear relationships occur in any system where two connected variables vary in a way which is proportional to each other. In particular many physical laws and relationships are linear, although the letters used will not always be x and y, and you must become flexible enough to deal with this. There are many examples to illustrate this; representing a relationship by a straight line usually comes from collected experimental data and the straight line is called a mathematical **model** of the data.

Example 2.9.1 The length L of a wire under load T satisfies the relationship $L = aT + b$ where a and b are constants. Plot the readings below on a graph and you should be able to estimate a and b by estimating the slope and the intercept on the L axis. What are the physical meanings of a and b?

Load T	50	100	150	200	250
Length L	10.7	11.8	12.8	13.95	15

(Answer: $a \approx 0.02$ is increase in length per additional unit load, $b \approx 9.63$ is unloaded length)

DO THIS NOW! 2.10

(a) The electric resistance R of an iron wire varies with the temperature t° C according to the formula $R = R_0(1+\alpha t)$ where R_0 and α are constants. Plot the following data values and hence estimate R_0 and α. What are the physical meanings of R_0 and α?

Temp t	20	40	60	80	100
Resistance R	22.8	25.6	28.35	31.25	34.1

(b) The volume V of a gas at constant pressure varies with temperature t° C according to the formula $V = V_0(1+\alpha t)$ where V_0 and a are constants. Plot the following data values and hence estimate V_0 and a. What are the physical meanings of V_0 and a?

Temp t	10	40	70	100	130
Volume V	50.8	56.25	161.7	67.1	72.6

(c) Fahrenheit (F) and Celsius (C) temperatures are related by:

$$F = \frac{9}{5}C + 32$$

Plot the graph of F against C (what is a sensible range?). What is the F intercept? What is the slope? Where does the graph cut the C axis?

2.10 STRATEGIES FOR HANDLING LINEAR EQUATIONS AND GRAPHS

There are several ways, algebraic, graphical, technological, of both thinking about and handling linear equations. By looking at the same idea from different directions, you can understand it better and can really convince yourself, and others, that your mathematical conclusions are correct.

Example 2.10.1 Describe at least three ways of solving $9x - 4 = 3x + 2$.

You could solve it completely algebraically; do some algebra to reduce it to standard form $y = mx + c$ and then find where it cuts the x-axis; plot and see where the straight line $y = 9x - 4$ meets the straight line $y = 3x + 2$ and this will ensure that the LHS = RHS; and so on.

Example 2.10.2 Here is an example of a more complicated equation for which a numerical solution is sought and for which several different approaches are taken.

Solve the equation $\quad \dfrac{x+3}{5} - \dfrac{2x+7}{3} = x$.

Example 2.10.2 Method 1: Pencil and Paper

Multiply by 15 (to remove fractions)	$3(x+3) - 5(2x+7) = 15x$.
Remove brackets	$3x + 9 - 10x - 35 = 15x$.
Subtract $15x$ and $(9 - 35)$ from both sides	$3x - 10x - 15x = -9 + 35$.
Tidy up	$-22x = 26$.
Finally	$x = -\dfrac{26}{22} = -\dfrac{13}{11} \approx -1.1818$ to 4 dp. .

Example 2.10.2 Method 2: Graphically

Plot $y = \dfrac{x+3}{5} - \dfrac{2x+7}{3}$ and $y = x$, and see where they cross, as illustrated in the graphic calculator output in Figure 2.3.

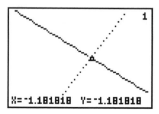

Figure 2.3 Solving a linear equation graphically.

You can read the crossover point more accurately by using any "Zoom" or "Find intersection" feature on your calculator. **Find out about these NOW**.

Example 2.10.2 Method 3: Numerically

You can tabulate the function values on, say, a graphic calculator tabulation function (e.g. Figure 2.4) or a spreadsheet, and simply read where the two functions are equal.

X	Y₁	Y₂
-1.182	-1.182	**ERROR**
-1.182	-1.182	-1.182
-1.182	-1.182	-1.182
-1.182	-1.182	-1.182
-1.182	-1.182	-1.182
-1.182	-1.182	-1.182
-1.182	-1.182	-1.182

$Y_2 = -1.818181818$

Figure 2.4 Solving a linear equation through tabulation.

Example 2.10.2 Method 4: Built-in features in Software Packages

Many CAS packages have built-in Solve features: this can provide a quick check of your working. For instance Figure 2.5 shows the TI-Interactive "solve" feature.

$$\text{solve}\left(\dfrac{x+3}{5} - \dfrac{2 \cdot x + 7}{3} = x,\, x\right) \qquad x = \dfrac{-13}{11}$$

Figure 2.5 Solving a linear equation using a typical built-in "Solve" function.

Example 2.10.2 Method 5: By a spreadsheet Solver (e.g. Microsoft Excel)

Spreadsheets are widely available and, amongst other useful purposes, a very useful verifier of solutions.

You can see some typical techniques in Figures 2.6(a) and 2.6 (b) on the next page.

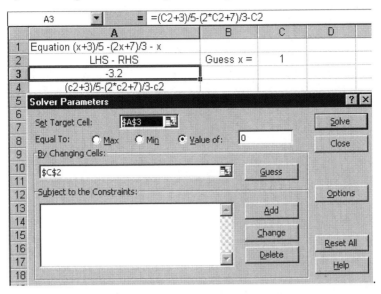

Figure 2.6(a) Solving a linear equation using the Excel "Solver" function:

the set-up using Solver from the Tools menu.

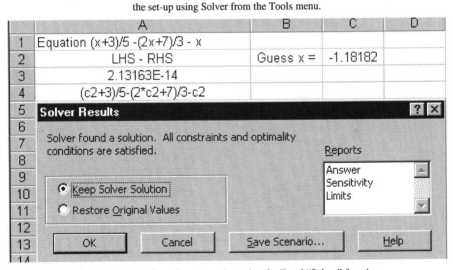

Figure 2.6(b) Solving a linear equation using the Excel "Solver" function:

the final result.

All these approaches then confirm that we have the "correct" answer. With this much power it is actually quite difficult to be wrong! Remember two things though.

- The final verification comes from putting the answer in the original expression but with "real" questions, where the answer has to be recorded to a certain number of decimal places, there is ALWAYS some small error involved.

- Remember to present your answer in a suitable form: Whom are you convincing?

DO THIS NOW! 2.11

Take these three examples (which you should have done, algebraically and otherwise, earlier in DO THESE NOW! 2.3) Solve them by a range of methods, algebraically, graphically and using technology, as in Example 2.10.2 above.

Remember you must be flexible about the letters used – your unknown variable plays the role of "x" even though it may not be called x – in fact it might really confusingly be called y!

(a) $4(y-3) = \dfrac{y}{2}+9$ (b) $2y+1 = \dfrac{y}{3}+6$

(c) $d+7 = 4\dfrac{(1-d)}{3}$.

END OF CHAPTER 2 - MIXED ACTIVITIES – DO THESE NOW!

Activity 2.1 More on linear functions

(a)   Determine the slope and y-intercept of the line $2x+3y-5 = 0$.

(b)   Find the equation of the line passing through (10, 0.4) which is parallel to the line $y = 2x-7$.

(c)   Find the equations of the lines shown in the graphical calculator screens in Figures 2.7(i) – (iv), and plot your equations to make sure they are right.

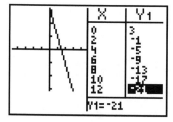

Figure 2.7(i)

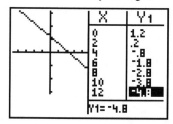

Figure 2.7(ii)

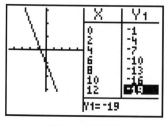

Figure 2.7(iii)

Figure 2.7(iv)

(d) Working both graphically and algebraically, find the value of x which satisfies the equation $5x-3 = 2x-7$.

Activity 2.2 More on formulae

(a) Make d the subject of the formula $E = \dfrac{p-d}{p}$. Evaluate d by hand when $E = \dfrac{7}{8}$

and $p = 3$. Check using your technological solver.

(b) Make n the subject of the formula $I = a + (n-1)d$.

How many ways can you think of to verify your answer? Do it!

(c) Rearrange the formula $\dfrac{1}{a} + \dfrac{1}{b} = \dfrac{1}{c}$ into each of the following forms

(i) $a = ?$ (ii) $b = ?$ (iii) $c = ?$.

Use your rearranged formulae to fill in the ? places in the following table, checking your answers technologically.

a	b	c
2	3	?
-4	?	3
?	1.3	4.7

(d) In a certain machine the applied effort E and the resistance overcome W are modelled by the equation $E = a + bW$, where a and b are constants. It is found by experiment that when $W = 50$, $E = 11.7$ and when $W = 150$, $E = 27.7$. Find a and b and find the value of W when $E = 15$.

(e)   Find a formula for w if $C = \dfrac{E + \sqrt{E^2 - 4Rw}}{2R}$.

Activity 2.3 Test your algebraic expertise

For (a)-(h), state ALWAYS TRUE or NOT ALWAYS TRUE, and *explain* why.

(a) $\dfrac{1}{x} + \dfrac{1}{y} = \dfrac{2}{x+y}$

(b) $\dfrac{1}{x} + \dfrac{1}{y} = \dfrac{x+y}{xy}$

(c) $\dfrac{x+1}{y+1} = \dfrac{x}{y} + \dfrac{1}{1}$

(d) $\dfrac{a}{b} + \dfrac{c}{d} = \dfrac{a+c}{b+d}$

(e) $\dfrac{x-2}{x+5} = \dfrac{-2}{5}$

(f) $\sqrt{x^2 + y^2} = x + y$

(g) $\dfrac{2a+5b}{5} = \dfrac{2a}{5} + b$

(h) $\sqrt{a^2 + b^2} = a + b$.

(Answers: F ,T, F, F, F, F, T, F)

3

Linking Algebra and Graphics 2

"An idea which can be used once is a trick. If it can be used more than once it becomes a method" George Polya

3.1 MORE ON CONNECTING ALGEBRA TO GRAPHS

In Chapter 2, when discussing the links between algebra and graphs, we restricted ourselves mainly to talking about linear functions – that is, those of the form $y = mx + c$.

This does indeed allow you to describe many, but by no means all, engineering systems. You now have the chance to go on and look at further mathematical functions which are important in engineering. Many of the general points about linking algebra and graphs will continue to be useful with these other functions.

To make our notation easier to extend, we shall start writing the linear expression $mx + c$ as $a_0 + a_1 x$, using the powerful subscript notation, with a_0 and a_1 representing constant quantities. To see why, we mention that the linear expression is an example of what is called a **polynomial expression** – which is any expression of the form:

$a_0 + a_1 x + a_2 x^2 + a_3 x^3 + a_4 x^4 ... + a_n x^n$. That is, it just contains terms which look like a constant times an integer power of x. The highest power present (n in this case) is called the **order** of the polynomial.

In this chapter we go on to look in some detail at polynomials of order 2, or quadratic polynomials (from the Latin *quadra* meaning "a square"). These are of fundamental importance in engineering, for instance in modelling suspension systems and electronic circuits, and the work in this chapter is an essential building block for several future chapters. So as usual, as you work through the chapter, just remember to keep on **doing all the exercises and activities**.

3.2 QUADRATIC FUNCTIONS

In the general subscript notation we would write a quadratic expression as $y = a_0 + a_1 x + a_2 x^2$, but many textbooks use a standard notation $y = ax^2 + bx + c$, so to match with what you have probably met before, we shall follow that for now.

The first useful thing to do with a quadratic expression $y = ax^2 + bx + c$ is to plot its graph for various values of a, b and c, to get a "feel" for what these parameters do to the graph. First look at the simplest quadratic graph $y = x^2$ in Figure 3.1.

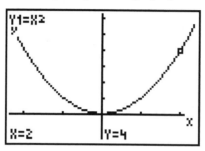

Figure 3.1 The simplest quadratic graph.

Note that adding the higher power term has made the graph curve in a particular way – it is no longer a straight line. Now go on to do these exercises to see what kind of features these functions have.

DO THIS NOW! 3.1

Plot the graph of $y = ax^2 + bx + c$ for various values of the parameters a, b and c as below (on your favourite technology). Observe how the graph changes as the parameters change and describe what you see in your own words.

In each case, fix the axes and scales:

(a) Plot together $y = x^2$, $y = x^2 - 1$ and $y = x^2 + 1$.

   Changing just c moves the quadratic how?

(b) Plot together $y = x^2$, $y = 2x^2$ and $y = 3x^2$.

   Changing just a does what to the graph?

(c) Plot together $y = x^2$ and $y = -x^2$.

   Making a negative does what to the graph?

(d) Plot together $y = x^2$, $y = x^2 + x$ and $y = x^2 + 2x$.

   Can you say anything simple about changing b?

You might have made the informal comments that changing c moves the graph up and down vertically; changing a changes the "sharpness" of the curve, and making a negative turns the graph upside down. Changing b shifts the curve sideways and downwards, but not in a simple way – more about this in chapter 5.

3.3 SOLVING QUADRATIC EQUATIONS
3.3.1 Connecting pictures and algebra with quadratics

As with straight lines, if you plot the graph of the quadratic curve $y = ax^2 + bx + c$ and look for the places where it cuts the x-axis, then those are the places where $y = 0$, that is, where $ax^2 + bx + c = 0$. That is, the values of x at those points give you the solutions or roots of the quadratic equation $ax^2 + bx + c = 0$.

These values of x are given by the formula $x = \dfrac{-b \pm \sqrt{b^2 - 4ac}}{2a}$, which you will almost certainly have met before. You can read about the derivation of this formula in section 3.4, but to explore a little more, work through the following exercise.

DO THIS NOW! 3.2

Plot the quadratic $y = ax^2 + bx + c$ for the values of a, b and c given below, and from your graphs and some calculation, fill in the table below where possible:

a	b	c	$b^2 - 4ac$	No. of times curve meets x-axis	$x = \dfrac{-b \pm \sqrt{b^2 - 4ac}}{2a}$
1	3	2	+1	2	-1 and -2
1	4	3			
1	1	0			
1	2	1			
1	0	0			
1	1	1			
4	4	1			

What do you observe about the relationship between the number of times the curve meets the x-axis and the value of $b^2 - 4ac$? Does the formula in the right-hand column indeed give the places where the curve meets the x-axis? Why do the sixth and seventh cases give difficulty - geometrically and algebraically?

Verify your conclusions by using other values of a, b, and c.

You should observe that the quantity $b^2 - 4ac$ is crucial in deciding the properties of both the curve and the solutions of the related quadratic equation.

- When $b^2 - 4ac > 0$, the quadratic cuts the x-axis twice.

- When $b^2 - 4ac = 0$, the quadratic cuts the x-axis once, although formally it "touches" the axis and cuts twice in the same place!

- When $b^2 - 4ac < 0$, the quadratic does not cut the x-axis. In this case you will find that the formula for the roots involves taking the square root of a negative number and so involves complex numbers (see chapter 6).

- The formula $x = \dfrac{-b \pm \sqrt{b^2 - 4ac}}{2a}$ does indeed give the place where the quadratic

graph cuts the x-axis – in other words, it gives the roots.

DO THIS NOW! 3.3

Your calculator, particularly if you have a graphic calculator, may well have a direct polynomial solver which will find the roots of a quadratic equation if you feed in the values of a, b and c. **FIND OUT NOW** if this is available on YOUR calculator! Find out how to use it and repeat the examples of DO IT NOW 3.2!

3.3.2 Connecting roots and factors with quadratics.

In the event that $b^2 - 4ac \geq 0$ then we can find so-called real roots or solutions of the quadratic equation. Let's label these α and β. We claim that it is then possible, and useful, to write the quadratic expression in the form: $ax^2 + bx + c = a(x - \alpha)(x - \beta)$ where $(x - \alpha)$ and $(x - \beta)$ are called the **linear factors** of the quadratic.

Earlier mathematics courses may have taught you the trick of finding these factors by inspection, but this is only of limited use in engineering, where the numbers are nearly always not simple and rarely whole. Using the formula to find the roots and then expressing the factors in terms of these, gives a foolproof way of finding the factors. To give you the confidence that this works, you could find the roots and factors and then simply multiply the brackets back out to see if you get back to the original quadratic, as in this example.

Example 3.3.1 Start with $y = x^2 + 5x + 6$.

- Plot it, and note where it cuts the x-axis, as in Figure 3.2.

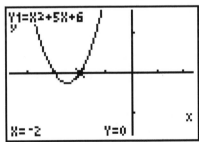

Figure 3.2 Finding the roots of a quadratic graphically.

- Now find its roots α and β, either using the formula or your built-in calculator function or preferably both. If you use both, make sure they give you the same answer! This should give $x = -2$ and $x = -3$ which agrees with the graph in Figure 3.2.

- Write down $a(x - \alpha)(x - \beta)$ and multiply out the brackets, and you should get back to the original quadratic.

$$a(x - \alpha)(x - \beta) = 1(x - (-2))(x - (-3)) = (x + 2)(x + 3) = x^2 + 5x + 6 \ .$$

DO THIS NOW! 3.4

Repeat Example 3.3.1 for these quadratic functions.

(i) $y = x^2 + 4x + 3$ (ii) $y = x^2 + 4x + 4$ (watch this special case)

(iii) $y = 1.23456x^2 + 3.45678x - 5.6789$.

3.4 AN ALGEBRAIC TRICK – COMPLETING THE SQUARE

There are various algebraic tricks associated with quadratic equations, some of which are made redundant by technology. One trick which may still prove useful is **completing the square**.

For example, the formula for finding the roots of a quadratic equation is derived by using this trick. It will also be useful at other times, for instance when dealing with Laplace transforms (see Chapter 17). Like so many algebraic tricks it is just a particular way of rewriting an algebraic expression, in this case a quadratic expression.

To see what it is about, consider $y = (x + A)^2 = x^2 + 2Ax + A^2$.

(check this out!)

This shows that any linear bracket squared can turn into a quadratic.

The "complete the square" method asks the reverse question:

Can you take any quadratic $y = ax^2 + bx + c$,

and write it as a bracket squared plus a constant $y = (x + A)^2 + B$?

That is we want to choose A and B so that
$$y = ax^2 + bx + c = (x + A)^2 + B$$

The answer is "Yes". First you see that a must be 1, but this can always be made the case just by taking out a factor of a (see DO THIS NOW 3.5!). Then, to see how the method works, it is easier to multiply out the bracket. Thus

$$y = ax^2 + bx + c = x^2 + 2Ax + A^2 + B$$

and by comparing coefficients of x^2, x and the constant, we get

$$2A = b \text{ or } A = \frac{b}{2} \qquad \text{giving } A$$

and $A^2 + B = c \text{ or } B = c - A^2$ giving B.

Example 3.4.1

Write the quadratic expression $y = x^2 + 4x + 9$ in the completed square form.

Check that it is correct by multiplying the brackets back out again.

In this quadratic: $a = 1$, $b = 4$ and $c = 9$.

Therefore $A = \dfrac{b}{2} = 2$ and $B = c - A^2 = 9 - 2^2 = 5$.

In completed square form then, $y = x^2 + 4x + 9 = (x+2)^2 + 5$.

Multiplying back out reveals that this has indeed worked:

$y = (x+2)^2 + 5 = x^2 + 4x + 4 + 5 = x^2 + 4x + 9$.

DO THIS NOW! 3.5

Complete the square with these quadratic expressions, verifying your answers:.

(a) $y = x^2 + 6x + 10$ (b) $y = x^2 - 4x + 12$

(c) $y = 2x^2 + 6x + 16$ (Hint - rewrite this one as $y = 2(x^2 + 3x + 8)$ and just complete the square on the bit in the bracket) .

We shall use this method again later but, just to finish off, here is an optional chance to derive for yourself the familiar formula for solving quadratic equations.

DO THIS! 3.6 (Not for the faint-hearted)
Derivation of the formula for solving a quadratic equation.
It is your job here to fill in all the steps.

Start with $ax^2 + bx + c = 0$

Divide by a (assuming $a \neq 0$) $x^2 + \dfrac{b}{a}x + \dfrac{c}{a} = 0$

Complete the square $(x + \dfrac{b}{2a})^2 + \dfrac{4ac - b^2}{4a^2} = 0$

Rearrange $x = \dfrac{-b \pm \sqrt{b^2 - 4ac}}{2a}$

Of course if you don't like all this algebra you can always get your CAS system to solve the equation – for instance Figure 3.3 shows a screen from a TI-92.

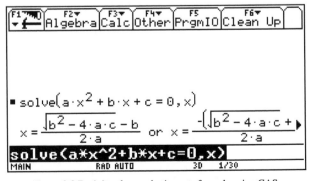

Figure 3.3 Deriving the quadratic roots formula using CAS.

3.5 A DIVERSION – MATCH THE GRAPHS WITH THE FUNCTIONS

In this section you have the chance to work on developing your mathematical "feel" by trying to match given algebraic quadratic equations with their graphs.

DO THIS NOW! 3.7

Below are 6 graphs and 12 quadratic expressions. Match each graph number with an equation letter. Verify your choice by plotting it.

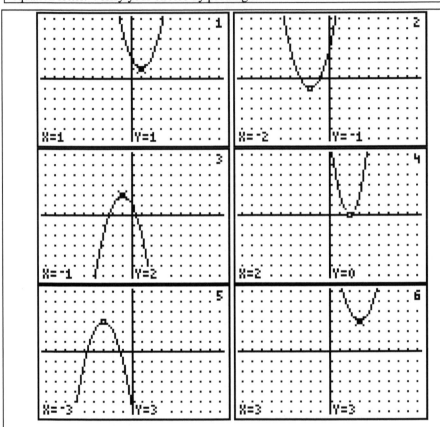

Figures 3.4 (1…6) Match expressions to these curves

(a) $y = x^2 - 6x + 12$ (b) $y = 0.25x^2 + 1.5x + 3$

(c) $y = x^2 - 12x + 13$ (d) $y = 0.25x^2 + 1.5x + 12$

(e) $y = -x^2 - 6x - 6$ (f) $y = -x^2 - 2x + 1$

(g) $y = x^2 - 12x + 13$ (h) $y = 0.25x^2 + 1.5x - 6$

(i) $y = x^2 - 2x + 2$ (j) $y = -x^2 + 4x - 6$

(k) $y = 2x^2 - 12x + 6$ (l) $y = x^2 + 4x + 3$

3.6 STRATEGIES FOR HANDLING QUADRATIC FUNCTIONS

To summarise, if linear models are not sufficient then we have to use a "curvy" model and the simplest "curve" we can use is the quadratic function. Hence it is necessary to be familiar with this type of expression. It has one unknown and contains powers up to the square:

$$y = a_2 x^2 + a_1 x + a_0 = ax^2 + bx + c .$$

The solution of quadratic equations forms a cornerstone in many engineering applications and in this section we are going to look at various methods of solution, summarising so far and introducing some new ideas.

The problem, then, is to solve $y = ax^2 + bx + c = 0 .$

We shall illustrate the various approaches and methods through examples.

Example 3.6.1 Solve the equation $x^2 - 5x + 6 = 0 .$

Example 3.6.1 Method 1: by finding factors

This is commonly the first method to be taught in schools, with "nice" whole number examples, but is of little use in engineering as the numbers are rarely "nice". Here is an outline, just to demonstrate the method and make links to what you may have done before, but it is not recommended.

We aim to rewrite the equation in the form $(x - \alpha)(x - \beta) = 0 .$

Multiplying out the brackets gives $x^2 - (\alpha + \beta)x + \alpha\beta = 0$, and comparing this with the original equation, we see that we must find α and β to satisfy

$$(\alpha + \beta) = 5 \text{ and } \alpha\beta = 6 .$$

By inspection, this requires $\alpha = 2$ and $\beta = 3 .$

Thus the quadratic equation can be rewritten as $(x - 2)(x - 3) = 0 .$

It is clear that either $x = 2$ or $x = 3$.

Example 3.6.1 Method 2: by formula

The roots of the quadratic are given by: $x = \dfrac{-b \pm \sqrt{b^2 - 4ac}}{2a}$.

For $x^2 - 5x + 6 = 0$, we have $a = 1$, $b = -5$ and $c = 6$

and so $x = \dfrac{-(-5) \pm \sqrt{(-5)^2 - 4 \times 1 \times 6}}{2 \times 1} = \dfrac{5 \pm \sqrt{25 - 24}}{2} = \dfrac{6}{2}$ or $\dfrac{4}{2} = 3$ or 2 as before.

Note the careful use of brackets here to help avoid getting the algebra wrong.

The two methods agree about the answer, but only *Method 2* is easily extendable to cases where the numbers are not whole and convenient.

Example 3.6.2 Solve the equation $x^2 - 5x - 8 = 0$.

Example 3.6.2 Method 3 by plotting the graph.

Figure 3.5 shows graphic calculator plots for $y = x^2 - 5x - 8$.

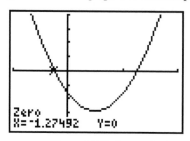 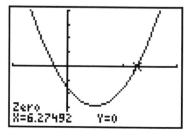

Figures 3.5 Finding quadratic roots graphically on a graphic calculator.

You can read from the graph, by "zooming in" as necessary, or by using any built-in features (in this case the "Zero" feature) that the roots are -1.27492 and 6.27492.

Example 3.6.2 Method 4: by using a spreadsheet such as Excel

The spreadsheet is more than an accounting tool. It can be useful for illustrating mathematical ideas and as a calculating tool. You can set it up to plot a graph as in Figure 3.6, to see that the roots exist. Then, from the Excel **Tools** menu, **Solver** or **Goal Seek** can be used to establish the values as accurately as necessary.

So we use the graph wizard to confirm that there are two roots near -1 and 6.

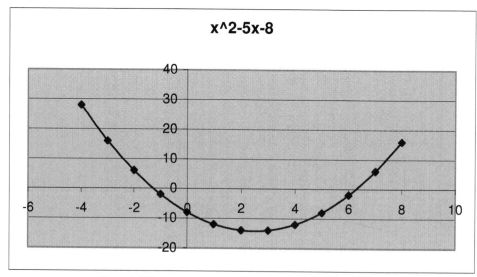

Figure 3.6 The result of plotting a quadratic graph in Excel

We can then use **Goal Seek** from the Tools menu as set up in Figure 3.7, to verify the roots obtained by Method 3 – but note that a good guess near each root is needed for the searcher to find the value.

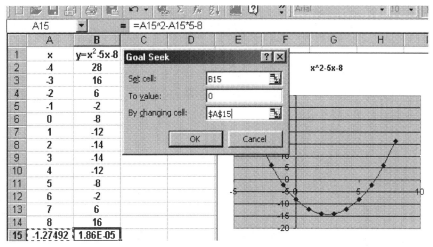

Figure 3.7 Refining the estimate of a quadratic root using Goal Seek in Excel.

Alternatively, we can use **Solver** from the Tools menu, as set up in Figure 3.8, but again a good first guess is required.

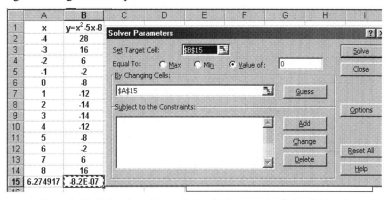

Figure 3.8 Refining the estimate of a quadratic root using Solver in Excel.

Example 3.6.2 Method 5: by customised software

Because solving quadratic equations is something an engineer must be able to do easily, there are many basic direct tools to do this. For instance, Figure 3.9 shows screen dumps from a graphic calculator showing such a package:

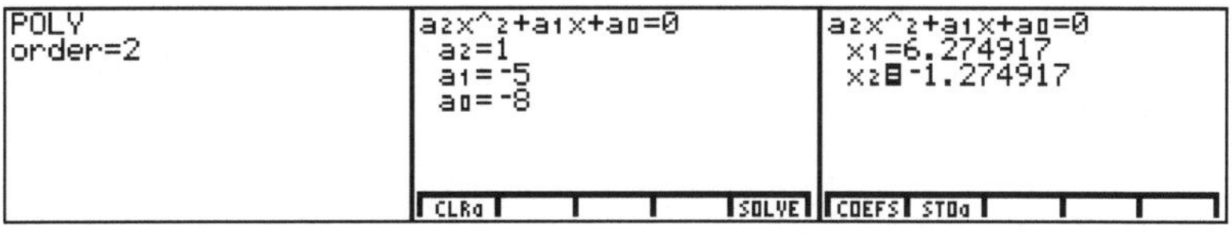

```
POLY           a2x^2+a1x+a0=0     a2x^2+a1x+a0=0
order=2        a2=1               x1=6.274917
               a1=-5              x2=-1.274917
               a0=-8



         CLRa                SOLVE COEFS STOa
```

Figure 3.9 Finding quadratic roots using a built-in graphic calculator function.

FIND OUT NOW what technology you have available to you for solving quadratic equations and make sure you know how to use it correctly.

3.7 WHERE NEXT WITH POLYNOMIALS?

Be warned: when you have powerful technology it can easily take you into new areas, sometimes before you are ready. Look at Figure 3.10 to see what happens if you use a graphic calculator to solve $x^2 + 2x + 5 = 0$:

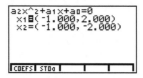

Figure 3.10 A graphic calculator finds complex roots of a quadratic equation

The notation is unfamiliar, since $b^2 - 4ac = 4 - 20 < 0$. There is no intersection with the x-axis, so solutions will not be real numbers. You are, in fact, meeting complex numbers – but you must wait for, or head for, Chapter 6 to find out more!

Finally, it is worth noting that polynomials generally are useful to approximate any function. For instance, the next most complicated in the polynomial family is the cubic $y = a_0 + a_1 x + a_2 x^2 + a_3 x^3$.

One of these is plotted in Figure 3.11 and you might observe that the extra term gives the possibility of two "bends" in the function. We shall deal with higher order polynomials in more detail in Chapter 18.

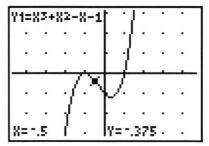

Figure 3.11 A cubic curve

END OF CHAPTER 3 ACTIVITY – DO THIS NOW!

The general quadratic equation is:- $ax^2 + bx + c = 0$

Solve it in each of the following cases by the formula method, verifying your results by plotting graphs and also by using as many methods and technologies as you have available to you.

a	b	c	Root x_1	Root x_2
5	30	44.2		
8	-54	63		
-0.7	0.53	0.6		
7	-23	5		

4

Other Essential Functions

"In certain problems that I have done it was necessary to continue the development of the picture as the method before the mathematics could really be done" Mark Kac

4.1 ESSENTIAL ENGINEERING FUNCTIONS

This chapter will allow you to run through the basic facts about some further mathematical functions which are fundamental to engineering - the exponential and logarithm functions, and the trigonometric functions. It will provide by no means a comprehensive view of these, but will allow you to encounter the main facts. These functions occur in many areas of engineering, so as you use them more and more, you will become more familiar with their behaviour and properties.

In informal descriptive terms, exponentials and logarithms describe systems which grow or decay; trigonometrical functions describe systems which oscillate, or have a periodic nature).

The key functions you will be handling are:

A general exponential function $\qquad y = a^x$.

The exponential function $\qquad y = e^x = \exp(x)$ where $e \approx 2.71828...$.

There will be more about why this number e is special in section 4.3.

The natural logarithm function $\qquad y = \ln(x) = \log_e(x)$.

$\qquad$ This is also called the logarithm to the base e.

The sine function $\qquad y = \sin(x)$.

The cosine function $\qquad y = \cos(x)$.

As you work through this chapter, just keep on **doing the exercises and activities**.

4.2 THE BASICS OF EXPONENTIALS AND LOGARITHMS
4.2.1 The number e and natural logarithms

DO THIS NOW! 4.1

Do this first and then read our comments below.

Plot the following functions on your favourite technology, over the range $-4 < x < 4$, thinking about what is a sensible range for y:

$$y = e^x = \exp(x) \qquad\qquad y = x \qquad\qquad y = \log_e x = \ln x$$

Comment on what symmetry you observe on the graphs. You should plot on equal scales on both axes – some calculators have a Zoom Zsquare mode for this - why?

You should have seen graphs something like those in Figure 4.1 below, which were plotted on a graphic calculator.

You must plot on equal scales in both directions, otherwise you will not see the symmetries - for example if you plot a circle on a scale of 1 unit per cm in the x-direction and 2 units per cm in the y-direction then it will not look like a circle, even though it still is one.

If you do this you will then see that the picture is symmetrical about the 45° line $y = x$. In other words, the graph of $y = e^x = \exp(x)$ is a reflection of the graph of $y = \log_e x = \ln x$ in this line of symmetry. Note that probably in your calculator the logarithm $\log_e x = \ln x$ and exponential $e^x = \exp(x)$ functions share a key, to reflect this close relationship.

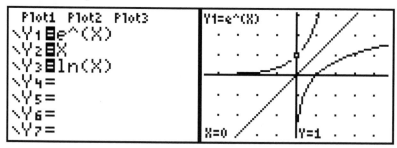

Figure 4.1 The symmetry in logarithm and exponential graphs.

Explore this more, by doing this next exercise, then reading our comments below it.

DO THIS NOW! 4.2

Evaluate on your calculator: $\log_e(e^2)$ $\log_e(e^3)$ $\log_e(e^4)$.

Predict the value of $\log_e(e^5)$? Check on your calculator ... are you right?

Evaluate on your technology: $e^{\ln 2}$ $e^{\ln 3}$ $e^{\ln 4}$.

What would you predict is the value of $e^{\ln 5}$? Check on your calculator. What general conclusion do you draw about the relationship between the logarithm and exponential functions?

You should see that $\log_e e^x = e^{\ln x} = x$ i.e. exponential and logarithm are inverse functions - one "undoes" the effect of the other. This explains the way the keys are set up on your technology.

Another way of writing this is that if $y = e^x$ then $x = \log_e y$ - an exponential on one side of an equation becomes a logarithm on the other.

Incidentally this gives meaning to the logarithm function - the logarithm to the base e (or natural logarithm) of a number y is just the *power* to which the base e must be raised to give that number y. So a logarithm should behave just like a power.

4.2.2. Logarithms to other bases

Most commonly as engineers and scientists you will use natural logarithms $y = \log_e x = \ln x$ - note the two ways of writing it. However you must also be aware that logarithms to other bases are also used.

Example 4.2.1 Common logarithms (base 10)

If $y = 10^x$ then $x = \log_{10} y$ is the power to which 10 must be raised to give y. This is the logarithm to base 10, or common logarithm. It is used in the definition of decibels and is also common because we have 10 fingers!

Example 4.2.2 Logarithms to base 2

If $y = 2^x$ then $x = \log_2 y$ is the power to which 2 must be raised to give y. It is used in information theory.

DO THIS NOW! 4.3

It is important to know how your machines and software represent these various exponential and logarithmic functions. For instance e^x is sometimes written exp(x) and in some software "e" has a special symbol (e.g. in Derive it is on a pad as $\hat{e}$.)

In Microsoft Excel for example (see Figure 4.2) the general log function has two parameters log(*number,base*), although LN and LOG10 are also there.

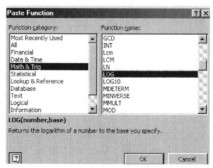

Figure 4.2 Finding logarithms amongst the standard functions in Excel.

Look NOW at your software and hardware and make sure you know how to work out the various logarithmic and exponential functions you need to use.

4.3 HOW THE EXPONENTIAL FUNCTION BEHAVES

4.3.1 The effect of k in $y = e^{kx}$ and $y = e^{-kx}$.

It is important for you to get a good feel for how the graph of the exponential function behaves - it is a very commonly occurring function in engineering, as you will see for instance in Chapter 15 in cooling or heating curves, radioactive decay, or electrical charge and discharge, and it has some interesting and useful properties.

DO THIS NOW! 4.4!

Do this first, then read our comments below.

Explore the GROWTH curve by changing $k > 0$ in $y = e^{kx}$.

Graph the following functions: $y = e^{x}$ $\qquad$ $y = e^{2x}$ $\qquad$ $y = e^{3x}$.

What would you predict $y = e^{4x}$ looks like? Plot it and see if you are right. Describe in your own words the effect on the graph of $y = e^{kx}$ of changing the parameter k.

Explore the DECAY curve by changing $k > 0$ in $y = e^{-kx}$

Graph the following functions: $y = e^{-x}$ $\qquad$ $y = e^{-2x}$ $\qquad$ $y = e^{-3x}$.

What would you predict $y = e^{-4x}$ looks like? Plot it and see if you are right. Describe in your own words the effect on the graph of $y = e^{-kx}$ of changing the parameter k.

Repeat the exercise above but with the function $y = 1 - e^{-kx}$.

You should have observed that, for $k > 0$, the positive exponential $y = e^{kx}$ models growth, as y increases with x, and that as k gets bigger, then the rate of growth increases. Correspondingly, the negative exponential $y = e^{-kx}$ models decay, as y decreases with x, and as k gets bigger, then the decay happens more quickly.

4.3.2 A magic property of $y = e^{x}$.

The function $y = e^{x}$ has a magic property. This is that at every point, the function has a value equal to the value of the slope of its graph at that point. This property is very important in the calculus, as you will see in various chapters later.

Recall that $e \approx 2.71828$. This is only an approximate value - it is an irrational number with an infinite number of non-repeating decimal places. You might wonder what is so special about this number that it should have magic properties, but there is no answer to that question - it is just the way things are!

Explore this property a little more now.

DO THIS NOW! 4.5

Given the two calculator screens in Figure 4.3, match up the functions with the curves. Describe what you observe about the set of curves $y = a^x$ as a increases.

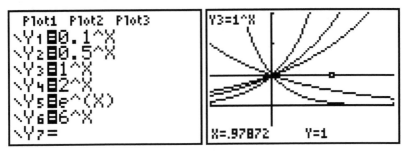

Figure 4.3 Match the functions to the curves

Try to reproduce these graphs on your own technology. In each case what is the value of y at $x = 0$? From your graphs estimate the slope of each curve at $x = 0$. Which has a slope closest to its function value at this point?

Answer all these questions **NOW** before reading on.

(You should observe that for $0 < a < 1$ the curves decay; when $a = 1$ the curve is just a horizontal line; and as a increases beyond 1, the curves increase to the right more and more quickly. All curves have a value of 1 when $x = 0$, but different slopes. It is curve Y_5, which is $y = e^x$, which has a slope of 1 when $x = 0$.)

Now concentrate on the curve $y = e^x$ (Y_5 in Figure 4.3). Working with your own graph on your technology, investigate the relationship between function value y and curve slope at other points. Figure 4.4 shows a few screens with the "tangent lines" drawn to help you with estimating the slope - work out the y value at the tangent point and look at the slope of the line.

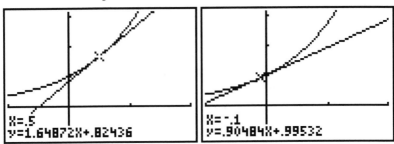

Figure 4.4

Finish doing this before reading on.

(You should find that for $y = e^x$ the value of y is the same as the value of the slope at *every* point - as we stated at the beginning of this section.)

4.4 HOW THE LOGARITHM FUNCTION BEHAVES
4.4.1 The laws of logarithms
A logarithm then is just a power, so logarithms and indices behave similarly.

There are **laws for logarithms** (which actually work for any base but which we present here for base e - natural logs), which correspond to the laws for indices, which you should have met before and with which you should become familiar:

Law 1: $\ln(AB) = \ln A + \ln B$.

Law 4: $\ln(\frac{1}{A}) = -\ln(A)$.

Law 2: $\ln(\frac{A}{B}) = \ln A - \ln B$.

Law 5: $\ln(1) = 0$.

Law 3: $\ln(A^n) = n \ln A$.

Law 6: $\ln(e^x) = x$.

DO THIS NOW! 4.6

In the following list, each of the expressions in the left hand column has an equivalent in the right hand column. Without using your calculator to evaluate the logarithms, match the pairs, stating which rule you are using. Only then, check using your calculator:

$\ln 3 + \ln 6$	$3 \ln 5$
$\ln 24$	$-\ln 8$
$\ln e$	0
$\ln 4 - \ln 16$	$\ln 6 + \ln 4$
$\ln 5^3$	$\ln 18$
$\ln 0.125$	1
$\ln 1$	$\ln 0.25$

4.4.2 Using the exponential/logarithm relationship
The inverse function relationship between $y = e^x$ and $x = \log_e y$ gives us a very useful algebraic tool. Here are some examples, of increasing difficulty, followed by some exercises for you to have a go at.

You will need to use a combination of ordinary algebraic skills and the laws of logarithms.

Example 4.4.1 Solve the equation $e^{-x} = 0.2$.

If $e^{-x} = 0.2$ then $-x = \log_e 0.2$, and so $x = -\log_e 0.2 \approx 1.609$.

Example 4.4.2 Solve the equation $e^{2x+1} = 0.2$.

If $e^{2x+1} = 0.2$ then $2x + 1 = \log_e 0.2$, and so $x = \frac{\log_e (0.2) - 1}{2} \approx 1.305$.

Example 4.4.3 Solve the equation $5^{x+1} = 0.1$.

If $5^{x+1} = 0.1$, then simply take natural logs of both sides.

This gives $\qquad\qquad\qquad\qquad\qquad\qquad \ln(5^{x+1}) = \ln(0.1)$.

Using Law 3 $\qquad\qquad\qquad\qquad\qquad\qquad (x+1)\ln(5) = \ln(0.1)$.

After a little manipulation: $\qquad\qquad\qquad x = \dfrac{\ln(0.1)}{\ln(5)} - 1 \approx -2.431$.

Example 4.4.4 Solve the equation $e^{-2x} = 3e^{-3.5x}$ and check the answer graphically.

Once more take natural logs of both sides $\qquad \ln(e^{-2x}) = \ln(3e^{-3.5x})$.

Using Law 1 and Law 6 $\qquad\qquad\qquad -2x = \ln(3) + \ln(e^{-3.5x})$.

Using Law 6 again $\qquad\qquad\qquad\qquad -2x = \ln(3) - 3.5x$.

Finally doing some algebra $\qquad\qquad\qquad x = \dfrac{\ln(3)}{1.5} \approx 0.732$.

The graphical check can be done, for instance, by plotting $y = e^{-2x}$ and $y = 3e^{-3.5x}$ and seeing where these two curves intersect, as shown in Figure 4.5.

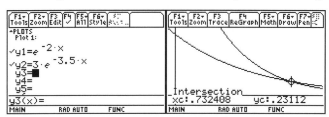

Figure 4.5 Solving an exponential equation graphically.

You can of course also check your solutions to Examples 4.4.1, 4.4.2 and 4.4.3 graphically! **DO THIS NOW.**

DO THIS NOW! 4.7

Solve these equations to 4 decimal places, first algebraically using logarithms and then confirming your results graphically.

(a) $e^x = 4$ $\qquad\qquad$ (b) $\quad e^{-x} = 0.1$ $\qquad$ (c) $\quad 4e^{3x-2} = 15$

(d) $e^x . e^{2x} = 3$ $\quad$ (e) $\quad 2^x = 5$ $\qquad\qquad$ (f) $\quad 5^{x+1} = 4$

(g) $2^{-x} = 5$ $\qquad\qquad$ (h) $\quad 5^{3x+2} = 4^{-2x}$ $\qquad$ (i) $\quad 0.423^x = 5.432$.

$\log_a b$ is the power to which base a is raised in order to get b. Use this to evaluate the following *without* using your calculator, then use the calculator to check.

(j) $\ln(e^3)$ $\quad$ (k) $\log_{10}(0.1)$ $\qquad$ (l) $\quad e^{\ln(5)}$ $\qquad$ (m) $\quad \log_{10} 1000$

Example 4.4.5 Some harder examples need a little imagination.

First solve graphically the equation $e^{-t} + e^{-2t} = 1$.

We get t to be about 0.48, by plotting $y = e^{-t} + e^{-2t}$ and seeing where $y = 1$. Do this yourself. What do you get? Do you agree?

Now can you see how you could solve this equation algebraically?

In fact you need to write it as a quadratic equation, with e^{-t} as the unknown - one set of mathematical skills building on top of others!

Begin with
$$e^{-t} + e^{-2t} = 1 .$$

Rewrite this as
$$e^{-2t} + e^{-t} - 1 = 0 .$$

Letting $x = e^{-t}$ and noting $(e^{-t})^2 = e^{-2t}$ $x^2 + x - 1 = 0$.

Solving this (by formula or whatever)
$$x = \frac{-1 \pm \sqrt{5}}{2} .$$

However since $x = e^{-t}$, and e^{-t} is always greater than 0 (look at its graph!), then we can only use the positive root
$$x = \frac{-1 + \sqrt{5}}{2} .$$

So
$$e^{-t} = \frac{-1 + \sqrt{5}}{2} \approx 0.62 .$$

This gives
$$t \approx -\ln(0.62) \approx 0.48 .$$

which verifies the graphical answer!

Finally, here are some activities which will prepare the way for the next section.

DO THIS NOW! 4.8

Find the solution t for each of (a) - (c), checking your answers both by substitution and also graphically by plotting the exponential function and seeing if your solution really does give the value on the left hand side:

(a) $0.6 = e^{-t}$ (b) $0.6 = 0.8e^{-2.5t}$ (c) $60 = 80e^{-0.029t}$.

4.4.3 The exponential function in action

In engineering, exponential decay occurs more often than growth. For instance, in movement, friction tends to wear down the motion. The decay graph $y = Ae^{-kx}$ has properties which are worth exploring, because of its common occurrence.

Example 4.4.6 Radioactive half-life

The level of redioactivity R in radioactive substances decays over time t, following a curve of the form $R = R_0 e^{-kt}$ where R_0 and k are constants which change from one substance and situation to another.

R_0 is the level of radioactivity when $t = 0$ - check this out by putting $t = 0$ and recalling from Chapter 1 that $e^0 = 1$.

k describes how quickly the radioactivity level decays - it is called the decay constant. Recall how this works by referring back to DO THIS NOW 4.5!

For Carbon 14, $R_0 = 6.53$ picocuries per gram, and k is $\dfrac{1}{8300}$.

DO THIS NOW! 4.9

Do these activities before reading our comments which follow.

Plot the graph of the radioactive decay curve for Carbon 14: $R = 6.53e^{-\frac{1}{8300}t}$.

From your graph read off how long it takes for the level of radioactivity to fall to 3.265, one half of its initial value.

Repeat this with any other point - that is, choose a starting value for t, read the value of R there and then record how long it takes for the level of radioactivity to fall to half of that value.

You should find that wherever you start, for a given substance with a given value of k, it takes the same length of time for the level to fall to one half of its starting value. This is called the radioactive half-life of the substance.

You can see that this will always be so by doing a little algebra as below - this shows you some of the power of algebra.

Starting with the decay curve $R = R_0e^{-kt}$, we ask the question how long it takes for R to fall to one half of its starting value R_0. In other words, what is t when $R = \dfrac{R_0}{2}$?

So we must solve for the value of t.
$$\frac{R_0}{2} = R_0e^{-kt} .$$

Divide both sides by R_0:
$$\frac{1}{2} = e^{-kt} .$$

Using Logarithm Law 6:
$$\ln(\tfrac{1}{2}) = \log_e(\tfrac{1}{2}) = -kt .$$

Finally:
$$t = -\frac{1}{k}\log_e(\tfrac{1}{2}) .$$

Note that R_0 has been cancelled out, so the length of time it takes for the level to halve is independent of the starting level - which you should have found before.

The half-life does however depend on k - different substances have different half-lives. Note that the half-life is NOT half the time it takes for the radioactivity to go completely!

Finish by checking, using this formula, that your graphical solutions were correct.

DO THIS NOW! 4.10

Try to get the answer to this puzzle without looking at the hints or the answer. Only look at them if you are really stuck!

The police discover a murder victim at 6:00a.m. The body temperature of the victim is measured then and found to be 25°C. A doctor arrives on the scene of the crime 40 minutes later and measures the body temperature again. It is found to be 22°C. The temperature of the room remains constant at 15°C. The doctor, knowing that normal body temperature is 37°C, is able to estimate the time of death of the victim.

If the cooling process is modelled by the equation $\theta = \theta_0 e^{-kt}$, where θ is the *excess* temperature over room temperature, estimate the time of death of the victim.

(Hints: Remember to work with θ as *excess temperature* above room temperature; work in minutes; take $t = 0$ to be at 6.00a.m. Use the two measurements taken to find the values of θ_0 and k. You can then identify the time of death when $\theta = (37-15)$, either by plotting the exponential curve and reading the value from the graph, or by solving the equation algebraically, or preferably both! Round off appropriately.)

(Answer: we made the time of death -88 minutes, or 4.32a.m.)

4.5 THE BASICS OF TRIGONOMETRIC FUNCTIONS
4.5.1 Extending the definitions to all angles

You saw in Section 1.6 how the sine, cosine and tangent of an angle are defined in terms of ratios of lengths of sides of a right-angled triangle. However this inevitably limited the definition to angles between 0° and 90° only.

In this section you will see how the definition can be extended to include all angles. This is done by imagining a point with co-ordinates (x, y), moving round a circle centred on the origin of radius r, as in Figure 4.6, with the angle θ being that measured from the positive x-axis as shown. The anti-clockwise sense creates a positive angle and the clockwise creates a negative angle. Trigonometric functions are often called **circular functions** for this reason.

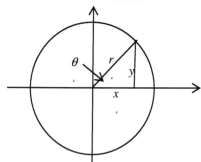

Figure 4.6 Defining the parameters for a general definition of trigonometric functions.

We thus re-define the trigonometric functions as

Sine	Cosine	Tangent
$\sin(\theta) = \dfrac{y}{r}$	$\cos(\theta) = \dfrac{x}{r}$	$\tan(\theta) = \dfrac{y}{x}$

4.5.2 The graphs of the sine and cosine functions

Having made definitions which can work for all angles, you now need to see how the graphs of the sine and cosine functions look and behave. You will see that they are useful for describing processes which oscillate or vibrate periodically - how many engineering situations can you think of where that happens?

DO THIS NOW! 4.11

Do all these activities and write your own comments before going on to read our comments which follow.

Plot the following sets of functions as specified over the range $-4 < x < 4$. Choose a sensible range for y. An important point is that you should now get used to working with your angles in **RADIANS**. Thus if you are using a calculator make sure your mode is set for this. You have already seen in Chapter 1 that 2π radians (just over 6.28) is equivalent to 360°; thus one radian is about 57°. There are good reasons for doing this – as you will see when we start discussing calculus in Chapter 7. Note that we sometimes use brackets to clarify expressions – this is a good habit.

(a) Plot together:

$y = \sin x$ $y = 2\sin x$ $y = 3\sin x$

Describe in your own words the effect of changing the parameter A (**amplitude**) in the function $y = A\sin x$.

(b) Plot together:

$y = \sin x$ $y = \sin(2x)$ $y = \sin(3x)$

Describe in your own words the effect of changing the parameter ω (**angular frequency**) in the function $y = \sin(\omega x)$.

(c) Plot together:

$y = \sin x$ $y = \sin(x+1)$ $y = \sin(x-0.5)$

Describe in your own words the effect of changing the parameter ϕ (**phase**) in the function $y = \sin(x+\phi)$.

You should observe that:

- The sine function forms a wave-like graph. Amplitude A changes the height of the sine wave so that $y = A\sin x$ varies between values of $+A$ and $-A$.

- A sine wave $y = \sin x$ repeats itself every 2π radians – this is called its **period** and is usually labelled as T. Angular frequency ω changes the speed of repetition of $y = \sin(\omega x)$ so that the period becomes $T = \dfrac{2\pi}{\omega}$.

- A sine wave "starts" (that is, it takes a value of 0 and is increasing) at $x = 0$. Phase ϕ moves the wave sideways so that $y = \sin(x+\phi)$ "starts" at $x = -\phi$.

DO THIS NOW! 4.12

(a) Repeat the exercises of DO THIS NOW 4.8! but with cosine instead of sine.

You should observe the same conclusions and also note that a cosine graph is identical to a sine graph, except that it is shifted to the left by a distance $x = -\dfrac{\pi}{2}$.

(b) Repeat also with the tangent function, $y = \tan x$.

You should observe that the same conclusions can be drawn again, but you should also spot that the tangent function repeats itself every π radians, and also becomes infinitely large at some values of x (e.g. $x = \dfrac{\pi}{2}$). Explain why this is so.

DO THIS NOW! 4.13

Here is a challenge – to see if you can make sense of the very rich pictures in Figure 4.7. If you can understand them (interpreting technological output is a useful skill) it may help you to consolidate your understanding of how trigonometric functions are linked to circles. Study the diagrams now – these comments may help.

You should see that there are three graphs.

- The first is a circle, where the point (x, y) which traces it out is given by $x = \cos(t)$ and $y = \sin(t)$, and t is the angle made by the line OP with the positive x-axis. On the technology this is $yt1$ against $xt1$.

- The second is a cosine wave $y = \cos(x)$, $yt2$ against $xt2$ on the technology.

- The third is a sine wave $y = \sin(x)$, $yt3$ against $xt3$ on the technology.

- The trace marker shows how the point (-0.5, 0.866025) on the circle corresponds to a cosine value of –0.5 and a sine value of 0.866025.

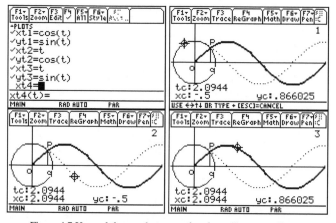

Figure 4.7 Unravel the puzzle to see what these diagrams tell you.

4.5.3 Some equivalences and symmetries with trigonometric functions

There are many relationships between and symmetries within the trigonometric functions. You should not expect to meet every one here but, as you find useful relationships, make a note of them.

DO IT NOW! 4.14

In the following, each of the functions in the left hand column is identical to one of the functions in the right hand column. By observing the graphs, match up the pairs:

$y = \sin(-x)$	$y = -\cos x$
$y = \tan(-x)$	$y = \cos x$
$y = \cos(-x)$	$y = -\sin x$
$y = \sin(x + \frac{\pi}{2})$	$y = -\tan x$
$y = \sin(x - \frac{\pi}{2})$	$y = \sin x$
$y = \sin(x + 2n\pi)$	(n is any integer (whole number))
$y = \cos(x + 2n\pi)$	
$y = \tan(x + n\pi)$	

Particular features you should notice include these:

- $\sin(x + 2n\pi) = \sin(x)$ expresses the fact that $y = \sin x$ repeats itself every 2π radians in the x direction; in other words if you increase x by any whole multiple of 2π, you get the same y value.

- $\cos(-x) = \cos(x)$ says that $y = \cos x$ is symmetric about the y-axis – going in the $-x$ direction gives you a perfect reflection of the shape going in the $+x$ direction. A function with this symmetry is called an **even function**.

- $\sin(-x) = -\sin(x)$ says that $y = \sin x$ is anti-symmetric about the y-axis – that is, going in the $-x$ direction gives you an upside-down reflection of the shape going in the $+x$ direction. A function with this symmetry is called an **odd function**.

In addition to these simple relationships there are many others, more complex. Sets of tables of these are readily available.

DO IT NOW! 4.15

Plot the graph of $y = \sin^2 x + \cos^2 x$; then plot the graph of $y = 1$. Deduce from your graphs that $\sin^2 x + \cos^2 x = 1$. Refer back to Chapter 1 to see the alternative justification of this by Pythagoras's theorem.

Use a similar graphical method to satisfy yourself of the validity of the following trigonometric identities; start building your own tables:

(a) $\sin(2x) = 2\sin x \cos x$.

(b) $\cos(2x) = \cos^2 x - \sin^2 x = 2\cos^2 x - 1 = 1 - 2\sin^2 x$.

Technology can help you to use these relationships, although you have to educate yourself as to how each tool handles them.

Example 4.5.1

A set of particularly useful relationships involves sines and cosines of sums and differences.

$$\sin(A+B) = \sin A \cos B + \cos A \sin B \qquad 2 \sin A \cos B = \sin(A+B) + \sin(A-B)$$

$$\sin(A-B) = \sin A \cos B - \cos A \sin B \qquad 2 \cos A \cos B = \cos(A+B) + \cos(A-B)$$

$$\cos(A+B) = \cos A \cos B - \sin A \sin B \qquad 2 \sin A \sin B = \cos(A-B) - \cos(A+B)$$

$$\cos(A-B) = \cos A \cos B + \sin A \sin B$$

You can find these in a CAS (see for instance Figure 4.8 for a TI-92 screen) but you can see there that it is not always obvious where they are.

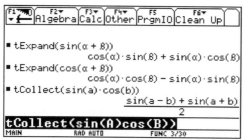

Figure 4.8 Trigonometric function formulae in a CAS system.

4.6 INVERSE FUNCTIONS AND TRIGONOMETRIC EQUATIONS
4.6.1 Inverse trigonometric functions

You are used to solving algebraic equations but now we are going on to look at solving equations which have trigonometric functions in them.

This will in the simplest cases lead to the idea of inverse trigonometric functions.

Example 4.6.1 Solve $\sin x = 0.5$.

A way of stating this problem in words is to say, "Find the angle whose sine is 0.5." You will be used to using your calculator to do this using the inverse sine button, which may be labelled as $\sin^{-1}()$ or perhaps as arcsin(), or even asin() in the Derive CAS. Note that the use of the notation $\sin^{-1}()$ must NOT be confused with raising to the power (-1) - the (-1) denotes the inverse function of sine - "the angle whose sine is" - rather than raising to a power.

Thus the answer is $x = \sin^{-1}(0.5) = \arcsin(0.5) \approx 0.524$,

where the value 0.524 (in radians, note) comes from your calculator.

However the graph of $y = \sin x$ reveals that things are not quite as simple as this. Take a look at Figure 4.9. It shows $y = \sin x$ plotted on the same graph as $y = 0.5$. It is clear that while the two graphs meet at $x \approx 0.524$, called the ***principal root***, they also meet at many other places - for instance $x \approx 2.618$, $x \approx 6.807$, and so on. In fact there are an infinitely large number of solutions. You should always be aware of this range of possible solutions.

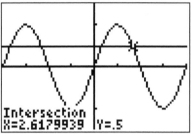

Figure 4.9 The graph shows multiple solutions of a trigonometric equation.

DO THIS NOW! 4.16

Text books often present the trigonometric function graphs with the x-axis marked in multiples of π (Figure 4.10). This usefully emphasises the symmetries in the waveform, but never forget that these are just numbers. Graphs made by technology often mark in numbers, so e.g. π will appear as approximately 3.142.

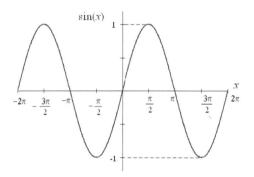

Figure 4.10 A graph of the sine function.

(a) Complete this table

°	0	30	45	60	90	180	270	360
radians (decimal)	0	0.524					3.142	
rad (multiple of π)	0	$\pi/6$					π	

(b) Find all solutions of $\sin x = 0.5$ over the range $-2\pi < x < 2\pi$ expressing your answers in degrees, in decimal radians, and in radians expressed as multiples of π.

Hint (only read this if desperate, or after you have produced your answers to part (b)): the first solution is 30° or $\pi/6$ or 0.524; using the symmetry from the picture, another is $\pi - \pi/6$, and so on.

Since cosine is a shifted version of sine, the next activity should not surprise you.

DO THIS NOW! 4.17

Repeat the process of Example 4.6.1, but this time with the equation $\cos x = 0.5$. This should generate the idea of the inverse cosine function $\cos^{-1}()$ or arccos() or acos() in Derive, to be understood as "the angle whose cosine is".

Principal root as given by calculator is $x = \cos^{-1}(0.5) = \arccos(0.5) \approx 1.047$, with other roots distributed according to the graph of $y = \cos x$.

Things are slightly different with the tangent function. Look at this.

DO THIS NOW! 4.18

Repeat the process of Example 4.6.1, but this time with the equation $\tan x = 1$. This generates the idea of the inverse tangent function $\tan^{-1}()$ or arctan() or atan() in Derive, to be understood as "the angle whose tangent is".

Principal root as given by calculator is $x = \tan^{-1}(1) = \arctan(1) = 0.785$, with other roots distributed according to the graph of $y = \tan x$, but note that this time the pattern of symmetries is different because you should recall that $y = \tan x$ repeats itself every π radians, and not only every 2π like $y = \cos x$ and $y = \sin x$.

4.6.2 More general equations involving trigonometric functions

In section 4.6.1 you met simple equations which led to the idea of the inverse trigonometric functions. In this section you will work with more complicated equations, but will still be using the same ideas.

In each case a graphical method - plotting aspects of the equation to see where it takes the right values - is indispensible for identifying where multiple solutions occur. Algebraic methods will make a contribution too.

Example 4.6.2

Find all solutions in the range $-2\pi < x < 2\pi$ of the equation $\sin(2x + \frac{\pi}{2}) = 0.6$.

To understand the multiple solutions, it is best to plot the graph of the function on the left hand side $y = \sin(2x + \frac{\pi}{2})$ over the given range (see Figure 4.11), and then read from the graph where that function takes the value $y = 0.6$. Values from the graph can be refined either by Zooming in, or by using any built-in solver to refine the value after a good first guess from the graph.

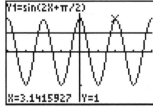

Figure 4.11 Looking for multiple solutions to a trigonometric equation.

Example 4.6.3

Find all the values x of arcsin(0.4) (that is, $\sin^{-1}(0.4)$) in the range $-4\pi < x < 4\pi$.

To solve this, plot $y = \sin x$ over the given range and identify the places where the graph takes the value $y = 0.4$. Note there are multiple answers.

END OF CHAPTER 4 - MIXED ACTIVITIES!

First, here are a couple more chances to work with the relationships between trigonometric functions; we have concentrated on sines and cosines as these are the most commonly used and you can always express problems in terms of them.

Activity 4.1

If $\sin(x) = \cos(x+a)$, what is the smallest positive value of a to make it true? What is the smallest negative value? (Thus even a cosine can always be turned into a sine!)

Activity 4.2

Find values of A, B and C so that an equation of the form $y = A\cos(Bx+C)$ corresponds to the graph in Figure 4.12. Each tick mark represents 1 unit. Check your answer by plotting it.

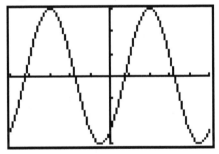

Figure 4.12 Hunt the function to fit the graph.

Activity 4.3

The graph of $y = (\cos(x))^2$ (sometimes written $y = \cos^2(x)$) is shown in Figure 4.13, in the viewing window $-\pi \le x \le \pi$ and $0 \le y \le 1$. Find an equation in the form $y = k + A\cos(Bx)$ that has the same graph and check by plotting it.

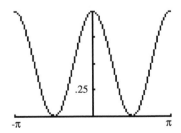

Figure 4.13 Hunt an alternative function to fit the graph.

Activity 4.4

Adding a sine and a cosine of the same angular frequency ω is a useful trick. It produces another *sinusoid* (sine wave) of the same angular frequency, but with different amplitude and phase. Work through this example.

Find R and α in the following:

$$R\sin(2x+\alpha) = 2\sin(2x)+3\cos(2x)$$

Expanding the left hand side with standard formulae (section 4.5) gives:

$$R\sin(2x)\cos(\alpha)+R\sin(\alpha)\cos(2x) = 2\sin(2x)+3\cos(2x) .$$

Making the $\sin(2x)$ and $\cos(2x)$ terms on both sides match:

$$R\sin(\alpha) = 3 \quad \text{(A4.3.1)} \qquad\qquad R\cos(\alpha) = 2 \quad \text{(A4.3.2)} .$$

We therefore need to solve these two equations for R and α. This is a standard process.

Square and add A4.3.1 and A4.3.2: $(R\sin(\alpha))^2 + (R\cos(\alpha))^2 = 3^2 + 2^2 .$

Simplify $R^2(\sin^2(\alpha)+\cos^2(\alpha)) = 13 .$

Recall that $\sin^2(\alpha)+\cos^2(\alpha) = 1$ $R^2 = 13 \text{ or } R = \sqrt{13} \approx 3.606 .$

Divide A4.3.1 by A 4.3.2 $\dfrac{R\sin(\alpha)}{R\cos(\alpha)} = \dfrac{3}{2} .$

Simplify $\tan(\alpha) = 1.5 \text{ or } \alpha = \arctan(1.5) \approx 0.983 .$

You can check your values by technology, either using built-in features (see for example Figure 4.14 which is a TI-92 screen), or by plotting both LHS and RHS and seeing that they match.

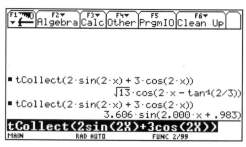

Figure 4.14 Using CAS to find trigonometric formulae.

Repeat this example but with $R\sin(x+\alpha) = 4\sin(x)+5\cos(x)$, and check your values for R and α by plotting both sides and seeing if the graphs are the same.

5

Combining and Applying Mathematical Tools

"concepts may be communicated easily in a format which combines visual, verbal, and symbolic representations in tight coordination" Scott Kim

5.1 USING YOUR TOOLBOX

In this chapter you have a chance to work some more with the basic ideas and tools you have met in Chapters 1-4, to try to understand them better. You will also have the chance to **use** those ideas to make a simple mathematical model.

5.2 THE MOST BASIC FUNCTION - THE STRAIGHT LINE

As an engineer, you are a keen observer of real physical situations and processes and must be interested in the ways in which mathematics can describe and help you to understand and predict those situations. It is extremely useful then to keep a set of pictures of your basic functions clearly in mind. It is a very valuable ability to "see" these functions, combine them simply together, and transform them, to make them fit and describe your situation.

Different situations are described by different functions, as you will see later in the book. Decay processes, such as radioactive decay, electronic discharge, or motion under friction, are frequently modelled by an exponential function like $y = Ae^{-kt}$. Oscillatory or vibratory processes are often modelled by a function such as a sine wave $y = A\sin(\omega t + \phi)$.

However, many real situations can be modelled simply by straight lines. We shall discuss this further in Chapter 18 when we deal with the Taylor series, but suffice it for now to say that if we look at any mathematical description of a system in close enough detail, every localized complex part appears linear. In Figure 5.1(a) you see a definitely non-linear function; in Figure 5.1(b) you see the area into which we are to zoom; and in Figure 5.1(c) you see the result of the zoom - which appears linear!

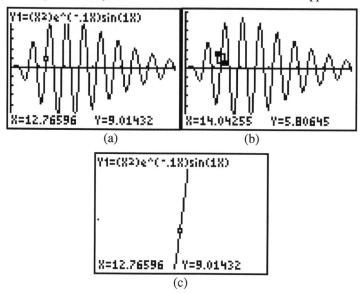

(a)   (b)

(c)

Figure 5.1 Zooming in to see local linear behaviour

This perhaps helps to explain why straight lines are regarded as so important in engineering mathematics (apart from the fact that they are easier to handle!).

5.3 TRANSFORMATIONS OF GRAPHS

Continuing the theme of connecting algebra to graphics, we now suggest you do some work to see how to move the standard graphs around by algebraically changing or transforming them.

5.3.1 The function notation

First you should make sure that you are comfortable with general function notation. We generalise this from functions you have already met, such as $y = mx + c$ or $y = \sin(x)$, to create the notation $y = f(x)$. The way to think about this is to consider f as an operation that changes elements that are put into it. For instance $f(x) = x^2 + 3$ is the operation "square the input which appears in the brackets after f and then add 3". Thus for instance $f(2x) = (2x)^2 + 3$, replacing every x by $2x$. Other examples appear in diagrammatic form in Figure 5.2.

The f function

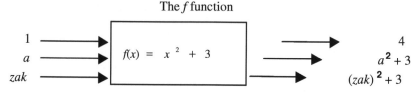

Figure 5.2 Illustrating the notion of a function.

DO THIS NOW! 5.1

If $g(x) = 2x^2 - 1$ and $h(x) = \sin(3x)$ write down NOW:

(a) $g(5)$ (b) $h(2)$ (c) $g(a)$ (d) $h(2x+1)$.

Answers: (a) 49   (b) -0.279   (c) $2a^2 - 1$ (d) $\sin(3(2x+1)) = \sin(6x+3)$

One important and useful skill is that of understanding the effect on the graph of a function, of changing the x in brackets to something else involving x, for instance $y = f(x) + a$, $y = f(x+a)$, or $y = f(ax)$.

To help you, technology such as a graphic calculator provides an easy way of drawing graphs quickly, and you should use your technology to develop your understanding and feel for the effect of algebraic transformations. A good way of dealing with this is not to go straight to plotting the graph, but to follow **POE**, an idea we borrow from Richard White and Richard Gunstone[1]:

> **P**redict what the outcome of an algebraic change will be.
> **O**bserve what actually happens by using the technology.
> **E**xplain the outcome.

You will often find your prediction will differ from your observed outcome, but these may be the situations where you learn most, as you have to explain what has gone wrong! Keep this in mind through the next sections.

5.3.2 Transformation of axes - vertical translations

To understand how to represent vertical translations of graphs algebraically, we ask the question: "Given a function $y(x)$ and its graph, what does $y(x) + a$, where a is a constant, look like,?"

Example 5.3.1

For the function $y(x) = x^2$ examine, using graphing technology, the graphical effect of the transformation $y(x) + a$ over a range of values of $a = $ -4, -3, ... , 3, 4.

Figure 5.3 shows the results of plotting the graphs on a graphing calculator.

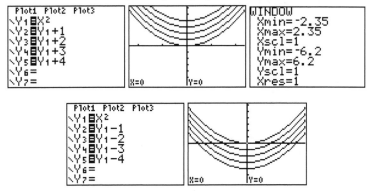

Figure 5.3 Vertical shift of a function graph

---

[1] Gunstone R and White R (1992) *Probing Understanding,* The Falmer Press.

You can see that what the algebraic transformation has done is to move the original curve $y(x) = x^2$ vertically by an amount a. If $a > 0$ it moves the curve upwards; $a < 0$ moves it downwards. In general this transformation of $y(x)$ to $y(x) + a$ is called a **translation** of the graph, parallel to the y-axis, of a units.

DO THIS NOW! 5.2

(a) Reproduce the graphs in Example 5.3.1. Change them to handle instead the function $y(x) = x^2 + 2x - 1$. Predict, using your own words, the effect of adding a constant to $y(x)$ to form $y(x) + a$, when $a = 8$. What happens when $a = -8$? Plot appropriate graphs to convince yourself that your prediction is a true statement.

(b) If you need further convincing, choose another function (e.g. $y(x) = \sin x$) and repeat the exercise with different values of a.

5.3.3 Transformation of axes - horizontal translations

Now we turn to how to describe horizontal translations of graphs algebraically. In this case we ask the question: "Given a function $y(x)$ and its graph, what does $y(x + a)$, where a is a constant, look like?"

Example 5.3.2

Once more we start with the simple function $y(x) = x^2$ and examine, using graphing technology, the graphical effect of the transformation $y(x + a)$ over a range of values of $a = -4, -3, \ldots, 3, 4$.

Figure 5.4 shows the results of plotting the graphs on a graphing calculator.

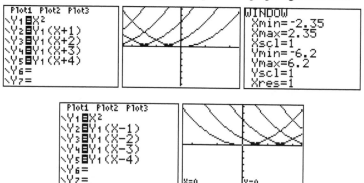

Figure 5.4 Horizontal shift of a function graph.

You can see that what the algebraic transformation has done is to move the original curve $y(x) = x^2$ horizontally by an amount a. If $a > 0$ it moves the curve to the left; $a < 0$ moves it to the right. Beware here, as some people feel that this is the "wrong way round", since they are used to right being positive! In general, the transformation of $y(x)$ to $y(x + a)$ is called a **translation** of the graph parallel to the x-axis of a units.

DO THIS NOW! 5.3

(a) Reproduce the graphs in Example 5.3.2. Change them to handle instead the function $y(x) = x^2 + 2x - 1$. Predict, using your own words, the effect of adding a constant to $y(x)$ to form $y(x+a)$, when $a = 6$. What happens when a = -6? Plot graphs to convince yourself that your prediction is true.

(b) Define a function $f(x) = \sin x$ and create its graph. Predict what the graph of $f(x+2\pi)$ looks like; then create it. What is happening here?

(c) If you need further convincing about any of this, choose another function (e.g. $y(x) = x^3$) and repeat the exercise with different values of a.

5.3.4 Stretching transformations

You have seen what happens when you *add* a constant to x or to y. Now what happens to the graph when you *multiply* x or y by a positive constant? We combine these together to ask the question: "Given a function $y(x)$ and its graph, what does $y(ax)$, where a is a positive constant, look like, and what does $ay(x)$ look like?"

Example 5.3.3

To examine $y(ax)$, take the function $y(x) = x^2$ again with values of a as 1, 2, 3, 4.

Figure 5.5 shows the results of plotting the graphs on a graphing calculator.

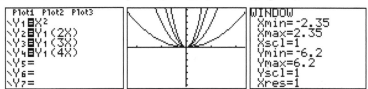

Figure 5.5 Horizontal "stretch".

You can see that what the algebraic transformation has done is to "squeeze" the original curve $y(x) = x^2$ inwards horizontally, with the amount of "squeeze" dependent upon a. For instance if $a = 2$, then the whole graph is squeezed inwards by a factor of $\frac{1}{2}$. The example may convince you that changing $y(x)$ to $y(ax)$ is a stretch, parallel to the x-axis by a factor $\frac{1}{a}$.

DO THIS NOW! 5.4

(a) Reproduce the graphs in Example 5.3.3. Change them to handle instead the function $y(x) = x^2 + 2x - 1$. Predict, using your own words, the effect of multiplying x by a constant a in $y(x)$ to form $y(ax)$, when $a = 6$. Predict what happens when $a = \frac{1}{2}$. Plot appropriate graphs to convince yourself that your prediction is a true statement.

(b) If you need further convincing about any of this, choose another function (e.g. $y(x) = x^3$) and repeat the exercise with different positive values of a.

Example 5.3.4

To examine $ay(x)$ take the function $y(x) = \sin x$ with values of a as 1, 2, 3, 4.

Figure 5.6 shows the results of plotting the graphs on a graphing calculator.

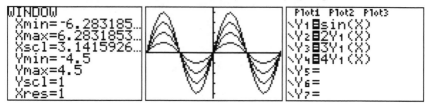

Figure 5.6 Vertical "stretch".

You can see that what the algebraic transformation has done is to "stretch" the original curve $y(x) = \sin x$ vertically, with the amount of "stretch" dependent upon a. For instance, if $a = 2$, then the whole graph doubles in height. In general, $ay(x)$ gives a stretch parallel to the y-axis by a factor of a.

DO THIS NOW! 5.5

(a) Reproduce the graphs in Example 5.3.4. Change them to handle instead the function $y(x) = x^3$. Predict, using your own words, the effect of multiplying y by a positive constant a in $y(x)$ to form $ay(x)$, when $a = 6$. Predict what happens when $a = \frac{1}{2}$. Plot appropriate graphs to convince yourself that your prediction is true.

(b) If you need further convincing about any of this, choose another function (e.g. $y(x) = \cos x$) and repeat the exercise with different positive values of a.

Why do you think we keep saying in these stretching cases that a must be positive?

5.3.5 Reflection of graphs

When we looked at the graphical effect of multiplying by a constant, we avoided negative constants. This is because allowing a negative constant introduces the complication of **reflection**. Here is the question: "Given a function $y(x)$ and its graph, what does the graph of $y(-x)$ look like?"

Example 5.3.5

To examine $y(-x)$, take the function $y(x) = x^3$ and plot both $y(x)$ and $y(-x)$.

Figure 5.7 shows the results of plotting the graphs on a graphing calculator.

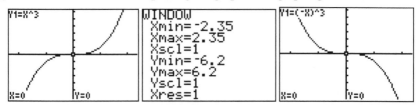

Figure 5.7 Reflection in the y-axis.

You can see that changing the sign of x has reflected the graph in the y-axis.

DO THIS NOW! 5.6

For the functions $f(x) = \sin x$ and $g(x) = \cos x$, predict the shapes of the graphs of $f(-x)$ and $g(-x)$. Plot graphs to convince yourself that your prediction is true. Say what you see in terms of *odd* and *even* functions, as identified in section 4.5.3.

5.4 DECAYING OSCILLATIONS

Many systems, for instance car suspensions, undergo motion which decays to rest or to a stable condition eventually, but the behaviour before reaching that often involves vibration or oscillation. It is therefore useful to have a mathematical function which describes this kind of behaviour and this is created by multiplying an oscillating function such as $y = A \sin \omega x$ by a decaying function such as $y = e^{-kx}$. You will find out when you study differential equations that this choice of function is not just lucky, but emerges as the right choice when certain assumptions are made about the dynamics of the system. More about that in Chapter 16.

Example 5.4.1 Plot the graph of $y = e^{-0.1x} \sin x$ for $x > 0$.

Figure 5.8(a) shows the basic shape of the graph. The range is chosen this way, since we can think of the x-axis as time, with movement starting at $x = 0$, and the y-axis as a displacement, where the oscillatory behaviour becomes clear.

Figure 5.8(b) shows the same graph, but with the graph of $y = e^{-0.1x}$ superimposed so it is clear that it is the exponential that is damping down the motion.

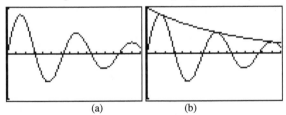

(a)   (b)

Figure 5.8 Representations of the decaying sine wave.

DO THIS NOW! 5.7

Plot on the same screen the graphs of $y = e^{-kx} \sin x$ for $k = 0.1, 0.2, 0.3$ and 1.

Write down, in your own words, the effect that k has on the graph of the mathematical function and on the moving system that the function describes.

5.5 A FOGGY FUNCTION

This is the first of two illustrations with which we finish this chapter, of how mathematics can be used to describe situations. They are different in style: this one concerns the use of a formula to find the reading age R of a piece of text. This is given by the **FOG INDEX**: $R = 0.4(\dfrac{A}{n} + \dfrac{100L}{A})$.

where A = number of words, n = number of sentences, and L = number of words with 3 or more syllables (excluding '-ing' and '-ed' endings).

DO IT NOW! 5.8

(a) Find four or five passages from a mixture of books, newspapers and magazines of varying reading difficulty. Using your common sense only, estimate the reading ages for each of these.

(b) Now work out the reading age of each using the **FOG** formula and compare with your estimate. Do you agree with the formula?

(c) Try to explain in your own words why you think the formula has been put together the way it has. For instance what do $\dfrac{A}{n}$ or $\dfrac{L}{A}$ represent?

Rearrange the equation so that you can find n given the other values. You could check your rearrangement here using a CAS technology, as shown for example with the TI-92 screen in Figure 5.9.

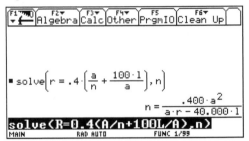

Figure 5.9 Changing the subject of the **FOG** index.

5.6 HEAT LOSS IN BUILDINGS - A MATHEMATICAL MODEL
5.6.1 Asking the question and building the model

In section 5.5, you met a formula which you were given and we asked you to explain why you thought it had been designed in the way it had.

In this section, you will approach an extended problem from a slightly different angle. You will take a situation and, with guidance, apply mathematical formulae to it, drawing conclusions about the artificial situation we present to you, and then having the opportunity to apply it to real life.

The situation concerns heat loss from buildings. In particular, we pose the question, "If you have a two-storey house, do you lose more energy if you only heat the downstairs, or if you heat both the downstairs and upstairs?"

The strategy is to partition the house into different zones and then to express the problem in terms of the heat flows from one zone to another through the various types of partition which separate them.

Do not underestimate the thought that lies behind the last sentence: how to set up a problem and how to express it in mathematical terms, is a skill you must work to develop. In this section, we will be leading you through that process, but at every stage you should be asking yourself why we suggest you do things the way we do. This is one stage of the process of mathematical modeling - a process which also involves a stage of making sense of things once you have "done the maths".

5.6.2 Setting up the variables and formulae

First take the main features of the situation and set them up in terms of algebraic variables, making any assumptions explicit.

Zones

For a two-storey house, use three zones: downstairs (d), upstairs (u) and outside (o). Label the temperatures in each zone as T_d, T_u and T_o (°C) respectively.

Label the temperature difference between zone a and zone b as $\delta T_{ab} = T_a - T_b$.

Label the rate of heat flow from zone a to zone b as q_{ab}. Thus, for instance, q_{uo} will be the heat flow from upstairs to outside (in watts W).

Partitions

Between any pair of zones, we shall assume:

- N is the number of types of partition materials (wall, window, etc), with each type being given a number $i = 1, 2, \dots$ etc;
- A_i is the area of partition type i (in m^2);
- U_i is a coefficient measuring insulation strength of partition type i (in Wm^{-2} °C^{-1}).

Assumptions

- The temperature in each zone will be constant and uniform.
- Each partition area - wall, window, etc - is of uniform construction.
- We treat doors as if they are walls.

The simplest model we can then use is that rate of heat flow will be proportional to partition area, to insulation coefficient (U value) and to temperature difference. A good principle in modelling is to start simply and see what insights that gives you.

For two given zones a and b, this then leads to the formula:

Total heat flow through all partitions between zones a and $b = q_{ab} = \displaystyle\sum_{i=1}^{N} A_i U_i \delta T_{ab}$.

DO THIS NOW! 5.9

The way we use mathematical notation can make a difference to how easy your written mathematics is for you and others to understand. Think about this.

We use *subscripts* such as i to label lists of similar quantities (e.g. partition number).

We use *double subscripts* such as du to make it clear that quantities (e.g. heat flow) depend on two other features (e.g. *from zone d* and *to zone u*).

We use the summation symbol Σ, a mathematical shorthand for writing down the sum of a list of similar expressions, ranging over the summation variable.

E.g. $\displaystyle\sum_{i=1}^{4} i = 1 + 2 + 3 + 4 = 10$ $\displaystyle\sum_{i=1}^{4} i^2 = 1^2 + 2^2 + 3^2 + 4^2 = 30$

$$q_{ab} = \sum_{i=1}^{N} A_i U_i \delta T_{ab} = (A_1 U_1 + A_2 U_2 + \dots + A_N U_N) \delta T_{ab}$$

5.6.3 Using the model for a particular house
In a particular house, the following detailed data are given:
- The house is 10m deep, 6m wide and each storey is 2.5m high.
- Each floor has a total window area of 8 m^2.
- The average external temperature is 5°C.
- When the heating is running in a given storey, a temperature of 20°C is maintained in that storey.
- The U value for the external walls of the house is $\quad$ 0.55 (Wm^{-2}°C^{-1})
- $\qquad$ " $\qquad$ windows $\qquad$ 2.9 (Wm^{-2}°C^{-1})
- $\qquad$ " $\qquad$ upstairs ceiling and roof $\qquad$ 0.34 (Wm^{-2}°C^{-1})
- $\qquad$ " $\qquad$ downstairs floor $\qquad$ 0.46 (Wm^{-2}°C^{-1})
- $\qquad$ " $\qquad$ downstairs ceiling and upstairs floor $\quad$ 1.0 (Wm^{-2}°C^{-1})

DO THIS NOW! 5.10

Use the data given to work through the steps below:

(a) Show that heat flow from upstairs to outside is $\qquad q_{uo} = 83.2(T_u - T_o)$.
(Walls, windows, ceiling/roof.)

(b) Show that heat flow from downstairs to upstairs is $\qquad q_{du} = 60(T_d - T_u)$.
(Ceiling/floor.)

(c) Show that heat flow from downstairs to outside is $\qquad q_{do} = 90.4(T_d - T_o)$.
(Walls, windows, floor.)

(d) Use part (a) to find the rate of loss of heat, in watts, from the upstairs through the walls, windows and roof, when the heating is *on* both upstairs and downstairs.

(e) Use the expressions you obtained in parts (a), (b) and (c) to find the temperature upstairs when the heating is on downstairs but not upstairs. (HINT: in a steady state, heat flow into upstairs = heat flow out of upstairs). Hence, find the rate of loss of heat from upstairs in these circumstances.

(f) Find the saving in the rate of loss of heat when the heating is on downstairs only, compared to when the heating is on both downstairs and upstairs.

(We get (d) 1248, (e) 11.28, 522.2, (f) 725.5.)

If you enjoyed this activity, you might like to extend it in various ways.

DO THIS NOW! 5.11

Do some or all of these:

(a) Set up a spreadsheet which will enable different configurations, areas, U values and temperatures to be investigated by just changing numbers in cells.

(b) Apply the above procedure to your own home environment and draw conclusions. Can you generalise the model by answering the question in algebraic terms only?

(c) How can the model be improved? (E.g. by including doors separately, etc.)

6

Complex Numbers

"In the real world, complex numbers don't exist." Dr. Phillip Christie

6.1 THE NEED FOR COMPLEX NUMBERS

All engineers need to know about complex numbers (which they sometimes refer to as j notation), why they are important, and how to handle them. If you would like to see one example of why, you might like to have a quick first look at Section 6.10, before reading each section and, most importantly, **doing the exercises**.

6.2 THE j NOTATION AND COMPLEX NUMBERS

The idea of a complex number springs from the seemingly peculiar step of seeking the solution to the equation

$$z^2 = -1. \tag{6.1}$$

Clearly this has no solution in the usual sense, since squaring any "real" number, positive or negative, leads to a positive answer. You might think that is that, but mathematicians made a bold step by supposing there is a solution and calling it i. Engineers call it j so as not to mix it up with electric current, so we shall use that, but be aware when using technology that mathematical software often uses i.

This seems to be a strange step to take but, surprisingly, what follows has real meaning and a deep impact in engineering, as you will see later.

In detail then j is defined as: $\qquad\qquad j = \sqrt{-1}$

so that squaring this $\qquad\qquad\qquad\quad j^2 = -1.$

Thus j is a solution to equation (6.1) as required. Because j is not a number in the way you normally understand, it is necessary to build an understanding of how this bit of mathematics behaves by looking at what happens in various circumstances. Begin by looking at what happens if you repeatedly multiply by j.

$$j^3 = j.j^2 = -j.$$

$$j^4 = j.j^3 = j.(-j) = -j^2 = -(-1) = 1.$$

Starting to multiply by j again just goes around the same cycle again: j, -1, $-j$, 1 and so on. To investigate these ideas further, it is useful to solve more quadratic equations.

Example 6.2.1   Solve $z^2 = -4$.

The solution (by hand – you may also try this on your technology) is

$$z = \pm\sqrt{-4} = \pm\sqrt{-1\times4} = \pm\sqrt{-1}\sqrt{4} = \pm2j.$$

You see that you do not need any additional strange steps: square roots of *any* negative number can be expressed in terms of j in this way.

Example 6.2.2   Solve $z^2 + 4z + 5 = 0$.

The solution (using the quadratic formula) is

$$z = \frac{-4\pm\sqrt{4^2 - 4\times1\times5}}{2\times1} = \frac{-4\pm\sqrt{-4}}{2} = \frac{-4\pm\sqrt{-1}\sqrt{4}}{2} = \frac{-4\pm j2}{2} = -2\pm j1 = -2\pm j.$$

Once more the only strange step needed is to use j. Again you may try to get your technology to do this: here it is on one graphic calculator:

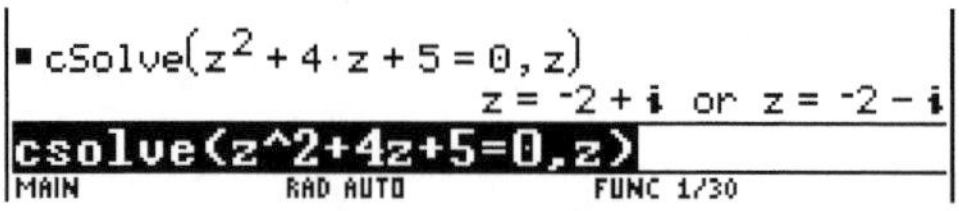

Figure 6.1 Solving a quadratic equation on a TI-89

(You will meet more of how these ideas interact with quadratic equations in Chapter 16 when you tackle differential equations.)

Note that the answers always come out in the form $z = x + jy = x + yj$; that is, one part not multiplied by j (x - the **real** part), and another part multiplied by j (y - the **imaginary** part).

A number such as z is called a **complex number.**

The **real part** of z is defined as $Re(z) = x = $ part not multiplied by j.

The **imaginary part** of z is $Im(z) = y = $ part multiplied by j (but *not* including j).

Thus for example $Re(3+j2) = 3$ and $Im(3+j2) = 2$ (note j is not included). A complex number for which $y = 0$ is called ***purely real***; one for which $x = 0$ is called ***purely imaginary***.

You should see that only one new step is needed – once you are allowed to use $j = \sqrt{-1}$, all these mathematical problems can be solved in terms of j. No other new steps are needed, although how you *think* about the new ideas is important.

6.3 ARITHMETIC WITH COMPLEX NUMBERS

All the normal rules of arithmetic and algebra apply. Just remember that

$$j^2 = -1.$$

Many calculators and computer packages will handle complex number arithmetic straightforwardly and you should find out about what *your* technology does.

To see how it works, it is best to look at some examples. Suppose then that there are two complex numbers $z = 1 + j2$ and $w = 2 - j1 = 2 - j$.

Then ***addition*** and ***subtraction*** are straightforward:

$$z + w = (1 + j2) + (2 - j1) = (1 + 2) + j(2 - 1) = 3 + j1 = 3 - j$$

$$z - w = (1 + j2) - (2 - j1) = 1 + j2 - 2 + j1 = (1 - 2) + j(2 + 1) = -1 + j3.$$

Multiplication is also straightforward if you remember about multiplying brackets out (see Chapter 2) and recall that $j^2 = -1$.

$$z.w = (1 + j2).(2 - j1) = 1 \times 2 + 1 \times (-j1) + (j2) \times 2 + (j2) \times (-j1)$$

$$= 2 - j1 + j4 - j^2.2 = 2 - j1 + j4 - (-1).2 = 4 + j3$$

Division is slightly more complicated by hand (although not with your technology of course!) because you want to write the answer in the form $x + jy$, so you need an algebraic trick to get the j off the bottom of the fraction. This trick involves the ***complex conjugate***.

The ***complex conjugate*** of a complex number $z = x + jy$ is defined as $\bar{z}$ or $z* = x - jy$ (that is, the conjugate is the same as z except that the imaginary part has had its sign changed). It has the special property that for *any* complex number, $z.\bar{z} = (x + jy)(x - jy) = x^2 + y^2 + 0j = x^2 + y^2$ which is purely real.

Now to sort out ***division*** of complex numbers, you just multiply the top and the bottom of your quotient by the complex conjugate of the bottom. Thus for example:

$$\frac{w}{z} = \frac{(2 - j1)}{(1 + j2)} = \frac{(2 - j1)(1 - j2)}{(1 + j2)(1 - j2)}$$

(multiplying top and bottom by conjugate of bottom),

$$= \frac{(2 - j1 - j4 + j^2 2)}{(1 + j2 - j2 - j^2 4)} = \frac{(0 - j5)}{(5)} = 0 - j1 = -j.$$

Note we have tidied up and taken care with signs; j has disappeared from the bottom as planned, and in this example the real part happens to be 0 so at the end we leave it out. Checking on the graphic calculator:

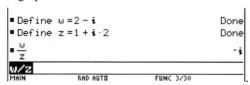

Figure 6.2 Complex arithmetic on a TI-89

DO THIS NOW! 6.1

Complex arithmetic on calculators:

Most recent calculators will handle complex arithmetic. **Find out now** how your technology handles complex numbers, concentrating for now on how to add, subtract, multiply, raise to a power and divide. Identify whether your technology uses "i" or "j" or some other form; check that you know how complex arithmetic works by doing the examples below. Work out how to find conjugates and real and imaginary parts. Talk to someone about it if you cannot sort it out.

Solve all of the following problems, both by hand and using appropriate technology, and thus check your own answers!

1. Solve the quadratic equations   (a) $x^2 = -9$ (b) $x^2 + 2x + 2 = 0$.

   In each case identify the real and imaginary parts of the roots.

2. Let $z = 1 + j$ and $w = -2 - 5j$. Calculate by hand and check by technology:

(a) $z + w$	(b) $z - w$
(c) $z.w$	(d) $\bar{z}$
(e) $\bar{w}$	(f) $z.\bar{z}$
(g) $2w - 3z$	(h) w/z
(i) $Re(z.w)$	(j) $Im(z.\bar{w})$

6.4 GEOMETRY WITH COMPLEX NUMBERS: THE ARGAND DIAGRAM

It is always useful to have a way of thinking about a mathematical idea through pictures. With complex numbers, the geometrical model for visualisation is called an *Argand diagram*. In this, the complex number $z = x + jy$ is represented by the point (x, y) in a two-dimensional plane, as in Figure 6.3.

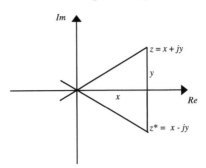

Figure 6.3 An Argand Diagram

The horizontal axis is called the **real axis**, as this measures x, the real part of z. The vertical axis is called the **imaginary axis**, as this measures y, the imaginary part of z.

The complex conjugate of z, z^* or $\bar{z}$, appears as the reflection of z in the real axis.

A complex number then is in a sense a two-dimensional quantity, represented by two real numbers in a certain order, x first, then y, i.e. an **ordered pair**. Indeed some graphic calculators reflect this by showing the complex number $x + jy$ as (x, y).

Example 6.4.1

Let $z = 1 + j$ and $w = -2 - 3j$. Observe the positions on the Argand diagram in Figure 6.4.

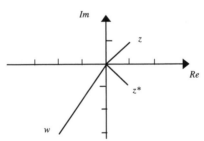

Figure 6.4 An example of an Argand diagram

DO THIS NOW! 6.2

Plot these complex numbers on an Argand diagram:

(i) $1 + 0j (= 1)$ (ii) $0 + 1j (= j)$ (iii) $-1 + 0j (= j^2)$

(iv) $0 - 1j (= j^3)$ (v) $1 + 0j (= j^4)$

What do you observe about the five numbers you have plotted? How would you describe what is done to the position of a complex number on the Argand diagram when you multiply by j? Think about and answer this before reading on into the following bracket.

(*You should get the idea that multiplying by j rotates the complex number anti-clockwise through 90°. Try it with some other examples of your own now.*)

    Dealing with a geometrical representation here leads to ideas for other ways of representing complex numbers and this is followed up in the next section.

6.5 CARTESIAN AND POLAR FORM, MODULUS AND ARGUMENT.

    The Argand diagram makes it clear that to each complex number there corresponds a unique point on the x-y plane. Thus any way we can define how to reach a point on that plane gives another way of representing the complex number. This leads to the idea of *polar form*.

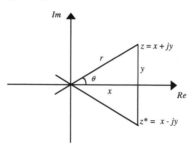

Figure 6.5 Polar form and the Argand diagram

    Look at Figure 6.5. The standard definition of $x + jy$ says "start at the origin, go x units in the real (Re) direction, then go y units in the imaginary (Im) direction". This is called the **Cartesian** or **rectangular form** of the complex number.

    An alternative way of reaching that same point is to say "start at the origin facing in the real direction, turn through an angle θ anti-clockwise, then go r units in that direction. This is known as the **polar form** of the complex number, sometimes written $r\angle\theta$, (r,θ) or $r\mathrm{cis}\theta$.

    Since these two forms represent the same number (and point), then there must be a way of going from one form to the other; this comes from the right angled triangle.

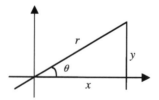

Figure 6.6 Geometrical connection between Cartesian and polar form

    From Figure 6.6, two relationships are apparent, using either Pythagoras' theorem, or the basic definitions of the trigonometric functions:

Converting from (r, θ) to (x, y)	Converting from (x, y) to (r, θ)
$x = r\cos(\theta)$ $y = r\sin(\theta)$	$r = \sqrt{x^2 + y^2}$ $\theta = \arctan\left(\dfrac{y}{x}\right)$ Beware the multiple values of the arctan function though - see section 4.6.1 and Example 6.5 below.

There is some jargon associated with this:

$r = \sqrt{x^2 + y^2}$ is the **modulus** of z, which can also be written $|z|$ or abs(z).

$\theta = \arctan\left(\dfrac{y}{x}\right)$ is the **argument** of **z**, which can also be written arg z or angle(z)

Since $x = r\cos\theta$ and $y = r\sin\theta$, then by replacing x and y in $z = x + jy$ there is a different way of writing z, as $z = r\cos\theta + jr\sin\theta = r(\cos\theta + j\sin\theta)$.

Summary: $z = x + jy = r(\cos\theta + j\sin\theta) = r\angle\theta$

Example 6.5.1 Convert the complex number $z = 1 + jl$ to polar form.

Applying the formulae:

$$r = \sqrt{1^2 + 1^2} = \sqrt{2} \approx 1.414.$$

$$\theta = \arctan\left(\frac{1}{1}\right) = \arctan(1) = 45° = \frac{\pi}{4}\, \text{radians} \approx 0.785\, \text{radians}.$$

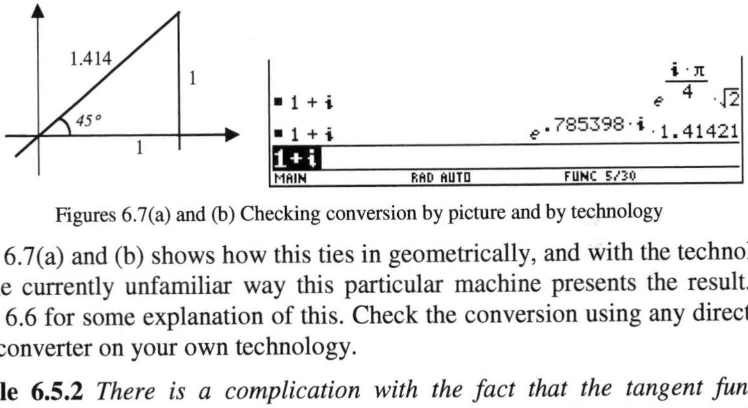

Figures 6.7(a) and (b) Checking conversion by picture and by technology

Figures 6.7(a) and (b) shows how this ties in geometrically, and with the technology. Note the currently unfamiliar way this particular machine presents the result. See Section 6.6 for some explanation of this. Check the conversion using any direct P-R or R-P converter on your own technology.

Example 6.5.2 *There is a complication with the fact that the tangent function repeats every 180°, so when using the arctan function you should also check that you are in the correct quadrant, as in this example.*

Convert the complex number $z = -1 - jl$ to polar form.

Applying the formulae:

$$r = \sqrt{(-1)^2 + (-1)^2} = \sqrt{2} \approx 1.414.$$

$$\theta = \arctan\left(\frac{-1}{-1}\right) = \arctan(1) = 45° = \frac{\pi}{4}\, \text{rads} = ????$$

Now on the face of it this looks like it is turning out exactly the same as Example 6.5.1, but that cannot be right since they are different numbers in different positions on the Argand diagram

To sort out what is happening, look at Figure 6.8:

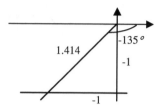

Figure 6.8 Watch your quadrant!

You should note that $\tan(45°) = 1$, but so also does $\tan(-135°) = 1$ (check on your calculator). Thus $\arctan(1)$ has two possible values, 45° and -135°, separated by 180°. To see which is the right one ALWAYS DRAW THE PICTURE TO CHECK.

As a back-up, you should see that the direct P-R or R-P converters or equivalent on your technology will sort this out automatically. In this case then the answer is shown and confirmed by technology in Figure 6.9.

$$\begin{cases} r \approx 1.414 \\ \theta = \arctan\left(\dfrac{-1}{-1}\right) = \arctan(1) = -135°\ = -\dfrac{3\pi}{4}^{c} \end{cases}$$

■ -1 - i	$e^{\frac{-3 \cdot i \cdot \pi}{4}} \cdot \sqrt{2}$
■ -1 - i	$e^{-2.35619 \cdot i} \cdot 1.41421$
-1-i	
MAIN RAD AUTO FUNC 9/30	

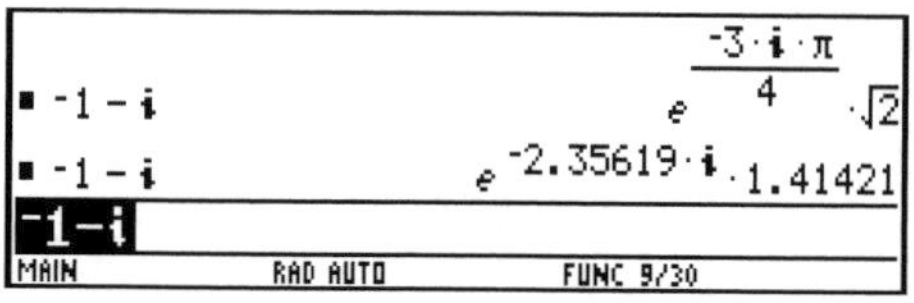

Figure 6.9 Checking the answer using technology – see Section 6.6 for explanation of notation here.

Example 6.5.3 Convert $5\angle 2.2°$ to rectangular form.

$$5\angle 2.2^{c} = 5(\cos(2.2) + j\sin(2.2)) \approx -2.94 + j4.04.$$

The multi-value complication does not occur this way round.

DO THIS NOW! 6.3

In this box, $z = 2 + j5$ and $w = -1 - j2$. In each of the following problems, do the calculations by hand and check by using the direct P-R or R-P converters or your own equivalent technology, and also by drawing Argand diagrams.

1. Evaluate: (i) $|z|$ (ii) $\arg(z)$.
2. Convert both z and w to polar form, either using the formulae, or using your technology, or preferably both, quoting the angles in both degrees and radians. Confirm that your solutions are correct by placing them on an Argand diagram and seeing if the diagram fits with your figures.

3. Convert to rectangular form (a) $2\angle 1.3°$ (b) $3\angle 35°$.

6.6 EULER'S RELATIONSHIP AND EXPONENTIAL FORM

One of the most astounding properties of complex numbers is **Euler's relationship**. This shows that, if you raise e to a complex power, then the exponential function is very closely related to trigonometric functions. This may surprise you, as you may be used to exponentials representing decay or growth, and sines and cosines representing oscillations. Nevertheless, without any proof, here it is - put it in your personal handbook.

> **Euler's relationship:** $e^{j\theta} = \cos\theta + j\sin\theta$ (*Note θ must be in **radians***)

This amazing result is surprisingly very useful in engineering mathematics, especially in analysing communication systems, and solving differential equations.

You will meet the justification of it in Chapter 18, but for now it is useful to become familiar with it, so here are one or two examples.

Example 6.6.1 $e^{j\frac{\pi}{4}} = \cos(\frac{\pi}{4}) + j\sin(\frac{\pi}{4}) = \frac{\sqrt{2}}{2} + j\frac{\sqrt{2}}{2} \approx 0.7071 + j0.7071.$

Example 6.6.2 Here is a very nice one: $e^{j\pi} = (\cos\pi + j\sin\pi) = (-1 + j0) = -1$.

In other words, combine π, which you can never know exactly, with e which you can never know exactly, with j, which is a figment of your imagination – and you end up with just the number -1. This is a remarkable result!

Exponential form

What Euler's relationship gives you is another very useful way of representing complex numbers. You have met already: $z = x + jy = r(\cos\theta + j\sin\theta)$

But from Euler's relationship, $\cos\theta + j\sin\theta = e^{j\theta}$, and so you can write z in the form

$$z = x + jy = r(\cos\theta + j\sin\theta) = re^{j\theta} \text{ - remember } \theta \text{ in radians!}$$

This is called the **exponential form** of the complex number z. You will note that it does not require any more calculation to find it. Once you have polar form then you have exponential form, as long as θ is in radians. This might help you to understand the notation appearing in the technology screens in Figures 6.7 and 6.9.

Example 6.6.3

$$z = 1 + j1 \approx 1.414(\cos 45° + j\sin 45°) \text{ from Example 6.5.1}$$

$$\approx 1.414(\cos\frac{\pi}{4} + j\sin\frac{\pi}{4}) = 1.414\angle\frac{\pi}{4}$$

$$= 1.414e^{j\frac{\pi}{4}} \text{ with the angle in radians.}$$

DO THIS NOW! 6.4

1. You have already converted $z = 2 + j5$ and $w = -1 - j2$ to polar form in "DO THIS NOW 6.3!". Convert z and w to exponential form.

2. Euler's relationship allows you to find more complicated complex exponentials. If $z = x + jy$, the result allows you to find

$$e^z = e^{x+jy} = e^x\, e^{jy} = e^x(\cos y + j\sin y).$$

Evaluate the following (by hand and/or by technology – how directly will your machine do this? Find out NOW):

(a) e^j (b) $e^{-1.1+3j}$

(c) $Re(e^{-1.1+3j})$ (d) $Im(e^{-1.1+3j})$.

3. If you plot $Re(e^{(-1.1+3j)x})$ you are plotting the graph of $y = e^{-1.1x}\cos(3x)$.

Satisfy yourself about this both algebraically and by plotting it on your technology.

4. Experiment with curves of the type

$$y = e^{ax}\cos(bx) \qquad \text{and} \quad y = e^{ax}\sin(bx),\ \text{as suggested below.}$$

Describe in words what happens when a is (a) positive, (b) 0, (c) negative.

Describe what happens when b is changed from small to large values.

Why do you think this type of function is useful in engineering? What kind of motion can it represent?

5. If $z = -0.1 + j3$ and t is an independent variable, use Euler's relationship to write down

(a) $Re(2e^{zt})$ (b) $Im(e^{2zt})$

Predict the shape of the graphs of both these functions for $t \geq 0$ and confirm by plotting them on your technology. Explain what you see in words.

If you use your technology to turn symbols into pictures, you can "see" the importance and use of Euler's relationship in engineering. Figures 6.10(a) and (b) show an example of how a complex exponential can represent damped oscillations:

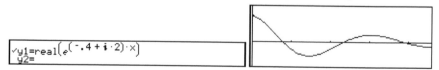

Figures 6.10(a) and (b) Complex exponentials represent damped oscillations

6.7 SOME USES OF POLAR AND EXPONENTIAL FORM

It turns out that it is particularly neat to handle multiplication, division and powers of complex numbers in polar form. Here is why. Have a look at these two complex numbers:

$$z_1 = r_1 \angle \theta_1 = r_1 e^{j\theta_1} \text{ and } z_2 = r_2 \angle \theta_2 = r_2 e^{j\theta_2}.$$

Then their **product** is (just using the laws of indices)

$$z_1 z_2 = (r_1 \angle \theta_1)(r_2 \angle \theta_2) = (r_1 e^{j\theta_1})(r_2 e^{j\theta_2}) = r_1 r_2 e^{j(\theta_1 + \theta_2)} = r_1 r_2 \angle (\theta_1 + \theta_2).$$

To *multiply* two complex numbers, *multiply* the moduli and *add* the arguments.

Their **quotient** is (again just using the laws of indices)

$$\frac{z_1}{z_2} = \frac{(r_1 \angle \theta_1)}{(r_2 \angle \theta_2)} = \frac{(r_1 e^{j\theta_1})}{(r_2 e^{j\theta_2})} = \frac{r_1}{r_2} e^{j(\theta_1 - \theta_2)} = \frac{r_1}{r_2} \angle (\theta_1 - \theta_2).$$

To *divide* two complex numbers, *divide* the moduli and *subtract* the arguments.

To raise to a power n (once more the laws of indices are needed):

$$z_1^{\ n} = (r_1 \angle \theta_1)^n = (r_1 e^{j\theta_1})^n = r_1^{\ n} e^{jn\theta_1} = r_1^{\ n} \angle n\theta_1.$$

To raise a complex number to the power n, raise the modulus to the power n and multiply the argument by n.

Example 6.7.1

Let $z = 3 - 2j$ and $w = 1 - j$. These give (check on your technology) in polar form

$$z \approx (3.606 \angle 0.588^c) \text{ and } w \approx (1.414 \angle -\frac{\pi}{4}^c).$$

Plotting these on an Argand diagram together with jz, $j^2 z$, $j^3 z$ and $z.w$

gives Figure 6.11:

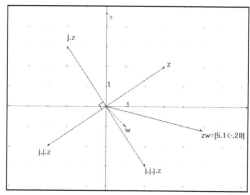

Figure 6.11 The picture shows the same story as the algebra

The multiplication by powers of j again illustrates the rotational aspect of j (90° in the anticlockwise direction) and zw shows the addition of angles $0.588 + (-\frac{\pi}{4})$ and multiplication of the moduli (3.606*1.414).

DO THESE NOW! 6.5

1. Convert the numbers 2 + 3j and 4 - 3j to polar form. **HENCE** evaluate the following:

(i) $(2+3j)(4-3j)$ (ii) $\dfrac{2+3j}{4-3j}$ (iii) $\left(\dfrac{2+3j}{4-3j}\right)^2$

Check your answers with your technology (Beware! Make sure you know and understand in what form they appear.)

Show the two original numbers on an Argand diagram. Add to the diagram your answers to (i), (ii) and (iii), *explaining* how the rules in polar form, particularly concerning adding and subtracting angles, show themselves on the diagram.

2. Evaluate

(a) $(2+3j)^3$ (b) $(0.1-0.2j)^9$

Check your answers here by using technology.

6.8 COMPLEX ALGEBRA

All of the above principles apply when you deal with complex *algebra* as opposed to just arithmetic, that is, with letters instead of or as well as numbers. The main differences are (a) you will have to do some algebra (!), and (b) you will need to use a CAS technology to check answers or to work out the more complicated cases.

Example 6.8.1

Figures 6.12(a) and (b) show a few simple algebraic examples, carried out on hand-held CAS technology, illustrating some basic rules. Can you identify them?

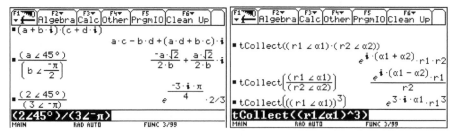

Figures 6.12(a) and (b) Complex algebra on a hand-held CAS system

---

DO THIS NOW! 6.6

Find out NOW how your CAS handles complex numbers. Test it out on the examples below, which as usual you should do on paper as well as on your machine.

1. Find the real and imaginary parts of z when

$$\frac{1}{z} = \frac{2}{3+2j} + \frac{1}{4-3j}$$

2. If $z = a + jb$ and $w = c + jd$ confirm by paper and pencil that your CAS is giving appropriate answers to the following:

$$z + w \quad z - w \quad 3z - 2w \quad zw \quad z\bar{z} \quad \frac{z}{w}$$

$$|z| \quad \arg(w) \quad w^2 \quad \frac{1}{(w+z)}$$

You may also like to substitute some numerical values in for a, b, c and d for checking purposes. Make sure you know about and can use the different modes of CAS output (rectangular, polar or exponential forms) available to you.

6.9 ROOTS OF COMPLEX NUMBERS
De Moivre'sTheorem

Note that $(\cos(\theta) + j\sin(\theta))^n = (e^{j\theta})^n = e^{jn\theta} = \cos(n\theta) + j\sin(n\theta)$. This result is known as *de Moivre's theorem.* This helps to clarify and inform the way to raise a complex number to a fractional power - for instance to take square roots:

$$z^n = r^n e^{jn\theta} = r^n(\cos(n\theta) + j\sin(n\theta)).$$

In fact it is a little more complicated than this, since sine and cosine have period 2π, so increasing θ by 2π will change nothing. Thus in general

$$z = re^{j\theta} = re^{j(\theta+2\pi k)} \text{ where } k \text{ is any integer } (0,\pm1,\pm2,\pm,3......), \text{ and so}$$

$$z^n = r^n e^{jn\theta} = r^n e^{jn\theta+2\pi kn}.$$

Normally the possibility of an extra 2π does not matter, but it does when n is a fraction, as it will indeed be when finding roots of complex numbers.

Example 6.9.1

Solve the complex equation $z^2 = -9$.

In general exponential form $z^2 = -9 = 9e^{j(\pi+2\pi k)}$.

Taking the square roots of both sides gives

$$z = (-9)^{\frac{1}{2}} = 9^{\frac{1}{2}} e^{j(\pi+2\pi k)/2} = 3e^{j(\pi+2\pi k)/2} = 3e^{j(\pi/2+\pi k)}.$$

Note that we take the positive square root of 9 here.

To interpret this, remember that k can take any integer values. All values of z have modulus 3 – that is, they lie on the circle centred on the origin of radius 3. Now at what angles do they lie? Taking $k = 0$ gives a z value at $\theta = \pi/2$. Taking $k = 1$ gives a z value at $\theta = 3\pi/2$. Taking further values of k simply repeats these two positions. There are thus just the two distinct values and just two square roots of –9. A diagram is a good check here since all the roots have the same modulus and in this case have an angle of $2\pi/2$ between their arguments.

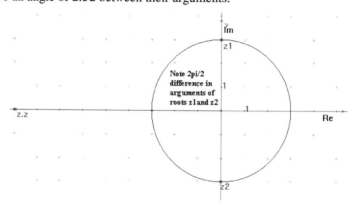

Figure 6.13 Roots are equally spaced around the circle

Example 6.9.2

Solve the equation $z^3 = 0.1 - 0.2\mathrm{j}$.

In exponential form we have $0.1 - 0.2\mathrm{j} = .223607e^{\mathrm{j}(1.10715 + 2\pi k)}$ so all solutions are given by $0.606962e^{\mathrm{j}(0.36905 + 2\pi k/3)}$ with $k = 0$, 1 and 2.

i.e. $0.606962e^{\mathrm{j}(0.36905)}$, $0.606962e^{\mathrm{j}(2.46345)}$, $0.606962e^{\mathrm{j}(4.455784)}$.

Note that in this case the roots are all separated by $2\pi/3$ (120°) and effectively cut the circle $|z| = 0.606962$ into 3 equal sections.

6.10 MINI CASE STUDY

Complex numbers occur, for instance, in two important areas of engineering: electronic circuit theory and differential equation theory. Because you meet more about differential equations in Chapter 16 onwards, we choose to present here a little case study about an electronic circuit.

An alternating current flowing in the circuit shown in Figure 6.14 is given by $I = I_0 \sin(\omega t)$ (ω is a constant and t is time). The corresponding voltage drops depend on ω and on the resistance R, capacitance C and inductance L of the circuit.

For a **resistance** R, the voltage V is $V = I_0 R \sin(\omega t)$ and this is "in phase" with the current.

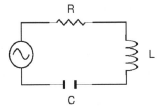

Figure 6.14 A simple series LCR circuit

For a **capacitance** C, the voltage $V = \dfrac{I_0}{\omega C} \sin(\omega t - \dfrac{\pi}{2})$, a phase "lag" of $\dfrac{\pi}{2}$ behind the current.

For an **inductance** L, the voltage $V = \omega L I_0 \sin(\omega t + \dfrac{\pi}{2})$, a phase "lead" of $\dfrac{\pi}{2}$ ahead of the current.

If the current $I = I_0 \sin(\omega t)$ is regarded as the imaginary part of a complex number $I = \mathrm{Im}(I_0 e^{j\omega t})$ then these various voltage drops can be represented very succinctly by recalling (refer back to Figure 6.11) that multiplication by j gives a lag of $\dfrac{\pi}{2}$ and division by j (which is the same as multiplication by –j) gives a lead of $\dfrac{\pi}{2}$.

Thus for **resistance** $V = RI$.

For **capacitance** $V = \dfrac{-jI}{\omega C}$.

For **inductance** $V = j\omega L I$.

In general if $V = IZ$ the factor Z is called the complex impedance, which is thus

$Z = R$ for a resistor, $Z = -\dfrac{j}{\omega C}$ for a capacitor, and $Z = j\omega L$ for an inductor.

END OF CHAPTER 6 - MIXED EXERCISES – DO ALL THESE NOW!
For each of the following, predict the answer by hand, check and observe by your technology, and explain and put right any discrepancies.

1. Let $z_1 = 2 - j$, $z_2 = -1 + 2j$, $z_3 = 3 + 4j$. Evaluate without using polar form, and present your answers in Cartesian form:

    (i) $z_1 + z_2 - z_3$ (ii) $z_1 z_2$

(iii) $\dfrac{z_1 z_2}{z_3}$ (iv) $(z_1 z_2)^2$.

2. Where it is an advantage to do so, repeat the calculations of question 1 using polar form, and present your answers in polar form.

3. If $z = -0.1 + 5j$ and t is an independent variable, use Euler's relationship to write down and plot the resulting function of t, for $t \geq 0$ in each of (i) and (ii):

   (i) $Re(e^{zt})$ (ii) $Im(2e^{zt})$.

4. The voltage E across an alternating current circuit is obtained by multiplying the total impedance, Z, of the circuit by the current I flowing in the circuit: i.e. $E = I.Z$.

   A certain circuit, in which a current $I = 4 + 5j$ flows, consists of three separate impedances, Z_1, Z_2 and Z_3 arranged so that the total impedance Z is given by

   $Z = Z_1 + (Z_2 Z_3)/(Z_2 + Z_3)$. Obtain the symbolic expression for the voltage E and calculate its value $(|E|)$, given

   (i) $Z_1 = 2 + 3j$ $Z_2 = 4 + 3j$ $Z_3 = 2 - 5j$

   (ii) $Z_1 = a + bj$ $Z_2 = 4 + 3j$ $Z_3 = 2 - 5j$.

5. Solve the equation $z^3 = 8 + j8$ and plot the results on an Argand diagram.

6. (Harder) Given $z = 4 + 5j$, write the following numbers in terms of z.

   $-9 + 40j$ $.09756 - .12195j$ $-236 - 115j$ 1.7921^c $.15617$.

7

Differential Calculus 1

Calculus is, alas, full of inert material." Peter Lax

7.1 THE NEED FOR DIFFERENTIAL CALCULUS

In any engineering system which actually or potentially changes, there will be a need to describe mathematically how quickly it changes – to find the rate of change of the quantities which describe the system. This is the subject of the differential calculus, and the process of finding rate of change is called differentiation.

You should understand the idea of a rate of change, interpreting it numerically and graphically. This will concern plotting a function and finding its slope. You will also use a symbolic approach which seems unrelated to what the process means!

To gain some idea of what it is that the process means, look at the next section, 7.2, where there are some short practical examples. These are simply presented there: they will be solved later, in Chapter 8. If you bring some previous experience of differential calculus with you, then you could try to solve them yourself before moving to chapter 8. Once more remember to **do the exercises and activities**.

7.2 DIFFERENTIAL CALCULUS IN USE

Differential calculus has many uses in describing the world. First here are several examples, NOT NECESSARILY FOR SOLVING NOW, but just to give some idea of the range of application. Having looked at them, you can go on to study the mathematical ideas here in Chapter 7, returning to these examples to solve them in Chapter 8. You should see how mathematics helps with understanding the processes and you should end up with a variety of ways to deal with such problems.

Example 7.2.1
A piece of A4 paper 29.7cm by 21.0cm is to have four identical square pieces with side of length x cm removed from the corners, so that it forms a template for an open-topped box. Find the dimensions of the box with the largest volume.

Example 7.2.2
(A similar, slightly harder situation.) An emergency petrol tank is designed to carry 1 gallon of petrol (4546 cm^3). Its shape can be considered a cuboid. The base of the cuboid is a rectangle with its length double its width. Find the dimensions of the tank that minimise the surface area required.

Example 7.2.3
A suspension system has displacement y given at time t by the function $y = e^{-0.5t} \sin(8.5t)$. What is its speed (rate of change of position or displacement) at time $t = 6$ seconds?

Example 7.2.4

In simple electronics you may have come across the relationship $I = \dfrac{Q}{t}$, where I is the current, Q is charge and t is time. This is only true for *constant* current. If the current varies with time, then the relationship is $i(t) = \dfrac{dq}{dt}$ which is the notation for the instantaneous rate of change of charge with respect to time.

In one circuit, the charge varies according to $q(t) = 0.1e^{-0.1t} \sin(5t)$. What is the maximum value reached by the current? What is the value of the current after a long time?

Now go ahead and explore the concept of differentiation.

7.3 WHAT DIFFERENTIATION MEANS GRAPHICALLY

7.3.1 Differentiation as finding the slope of a curve
This section provides first a classical description of the graphical interpretation of differentiation and suggests some graphically based activities which illustrate and expand the basic ideas of the differential calculus.

The process of differentiation is about finding the rate at which a function is changing. In graphical terms, this is represented by the slope of a graph of the function. This is where you will meet the mathematical ideas; later you will find these ideas applied in engineering situations.

Geometrically, the process of differentiation gives the slope of a curve at any point. In fact, to be more precise, it gives the slope of the line which is the tangent to the curve at each point. Algebraically, the process of differentiation of a function $y = f(x)$ with respect to the variable x is defined mathematically as:

$$\frac{dy}{dx} = \underset{\delta x \to 0}{Limit} \left(\frac{y(x + \delta x) - y(x)}{\delta x} \right),$$

but this formula needs some interpretation, first with the notation and terminology.

$\dfrac{dy}{dx}$ is called the *derivative*, or *derived function* of $y(x)$ with respect to x. It gives the

slope of $y(x)$ at any point x and looks like it is something to do with $\dfrac{change\ in\ y}{change\ in\ x}$.

7.3.2 Exploring graphical differentiation through an example

Perhaps it is worth looking at a simple example of a function, drawing its graph, and investigating it to see what the definition of differentiation means in terms of a picture.

Mathematically the simple function usually chosen in books is $y = x^2$ and we shall not break that tradition!

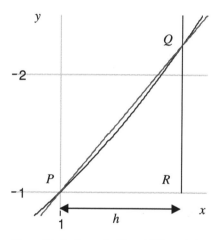

Figure 7.1 Getting at the idea of differentiation

To calculate a slope you need the change in y and the change in x, so consider two points a small distance h apart: P is the point $(1, 1^2)$ and Q the point $((1 + h),\ (1 + h)^2)$, where h is the distance from P to the point R, that is, it is the change in x from P to Q. The change in y from P to Q is correspondingly given by $((1 + h)^2 - 1^2)$. You can then find the slope of the line PQ by dividing the change in y by the change in x.

Algebraically you can express this as

$$\text{Slope} \approx \frac{y(x+h) - y(h)}{h} = \frac{(1+h)^2 - 1^2}{h}.$$

The "approximately equals" sign is because, as you can see from Figure 7.1, the slope of PQ is not exactly equal to the slope of the curve at P. However, the smaller h is, the closer the slope of PQ will be to the slope of the curve at P. The definition of the derivative can then be visualized as the limit of the slope of the line PQ as Q slides down the curve towards P – that is as the value of h nears 0.

DO THIS NOW! 7.1

Explore this idea of a *limiting process* using technology: (graphic calculator or spreadsheet?). The expression for the slope, or value of the derivative, of $y(x)$ is

$$\frac{dy}{dx} = \underset{h \to 0}{Lim}\left(\frac{y(x+h) - y(x)}{h}\right) = \underset{h \to 0}{Lim}\left(\frac{(1+h)^2 - 1^2}{h}\right). \quad \text{Evaluate} \quad \left(\frac{(1+h)^2 - 1^2}{h}\right) \quad \text{for}$$

various values of h, for instance $h = 1$, then 0.1, then 0.01, 0.001, 0.0001, 0.00001, etc. In other words, let $h \to 0$ and see what value the expression approaches.

Why can't you take it as far as actually calculating for $h = 0$?

You should find that the value of the expression should tend towards 2 as h gets smaller. Thus the slope of the curve $y = x^2$ at $x = 1$ is actually 2.

Figure 7.2 shows a quite neat way of doing this on a graphic calculator (in this case a TI-83-Plus using the Table key).

Figure 7.2 Approaching differentiation via a table of numbers.

DO THIS NOW! 7.2

Repeat the limiting exercise for $y = x^2$ but this time at $x = 2$. That is, estimate the

value of $\frac{dy}{dx} = \underset{h \to 0}{Lim}\left(\frac{(2+h)^2 - 2^2}{h}\right)$. You should get an answer of 4. Repeat for

various other values of x, say $\{-1, 0, 2, 3....\}$

Plot the graph of y against x. Plot also your calculated values of $\frac{dy}{dx}$ against x.

Hence can you guess a formula for the derivative at any value of x?

You should guess that the formula giving $\frac{dy}{dx}$ when $y = x^2$ is $\frac{dy}{dx} = 2x$. Thus your

plotted points should lie on the line $y = 2x$. You can confirm this using one of your technologies. For instance Figure 7.3 shows it on a graphic calculator.

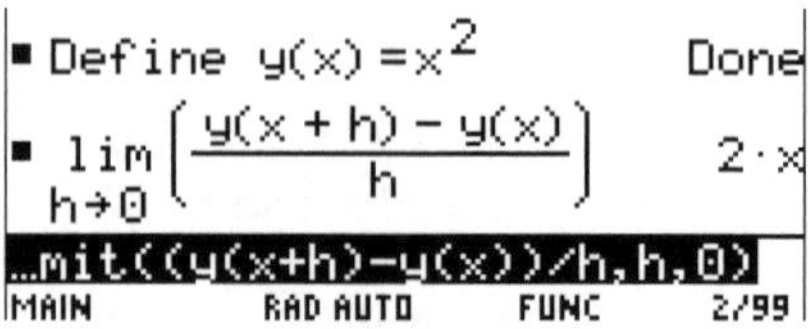

Figure 7.3 Using CAS to confirm a differentiation formula

There is an algebraic way of showing that this is the case; you will meet with that approach in section 7.6.

When you plot both the function $y = x^2$ and its derivative $y = 2x$ on the same axes, as in Figure 7.4, you can make sense of the picture as follows:

- When $x = 0$, the function $y = x^2$ is horizontal and so has zero slope. Sure enough, at this point the derived function has a value of 0.

- As x becomes more positive, the function $y = x^2$ has an increasingly positive slope, and the derived function has an increasingly positive value.

- As x becomes more negative, the function $y = x^2$ has an increasingly negative slope, and the derived function has an increasingly negative value.

Thus the derived function sensibly describes the variation in the slope of the curve.

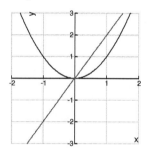

Figure 7.4 A function and its derived function plotted on the same axes.

Explore this approach to the idea further by doing the next example.

DO THIS NOW! 7.3

Plot, by any technological means you choose, the function
$$y = \sin(x) \quad (\text{Range } -2\pi < x < 2\pi)$$

Note from the graph the place(s) where the slope is 0, where it is at its largest positive value, and where it is at its largest negative value.

Hence predict roughly what the graph of $\dfrac{dy}{dx}$ against x for this function will look like. Do you find it possible to put rough values on this graph yet? If so, how? Check your conclusions by plotting the graphs as follows either:

(a) on a graphic calculator, plotting $y = \sin(x)$ and using any built-in *numerical* differentiation feature, such as NDERIV on the TI-83-Plus, or

(b) using the algebraic knowledge that $\dfrac{dy}{dx} = \cos(x)$, or otherwise.

7.4 VARIOUS WAYS OF FINDING DERIVATIVES

Graphical calculators mostly use a numerical approach to differentiation, estimating the slope as you have been doing. Those which do not have a CAS (Computer Algebra System) will present the results numerically or graphically. Further exploration of this idea follows in section 7.5.

There are pencil and paper algebraic techniques for finding a formula for the rate of change of a function - the slope of its graph at any point. This is the traditional emphasis in engineering mathematics courses and is tackled in section 7.6.

The technologically supported way to find a derivative is to use a PC-based CAS, such as Derive, Mathcad, Maple, Mathematica, or TI-Interactive! or hand-held machines such as the TI-89, Voyage 200, or Casio ALGEFX2.0PLS. You can use these to build up a table of standard derivatives of your own, to keep in your personal handbook, and you can think of symbol manipulators as a large and flexible look-up table. Some discussion of these appears in section 7.7.

7.5 NUMERICAL DIFFERENTIATION

The rate of change of a function $y(x)$ is approximately $\dfrac{change\ in\ y}{change\ in\ x}$, with a limiting process being needed if the rate of change is not constant. A graphical calculator uses numerical differentiation to calculate the derivative curve, as in Figure 7.5:

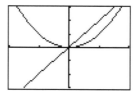

Figure 7.5 A graphical calculator screenshot showing a function and its derivative.

To find the gradient at (x_1, y_1), then $\dfrac{change\ in\ y}{change\ in\ x} \approx \dfrac{y_2 - y_1}{x_2 - x_1}$.

The calculator will then repeat this with the point (x_2, y_2), which lies on the curve to the right of (x_1, y_1), moving closer to (x_1, y_1) until there is less than a specified amount of change in the calculated value of the gradient. This is then repeated on every point of the curve to build up the graph of the gradient.

In the next activity we suggest you make use of a spreadsheet - we have used Microsoft Excel - to explore numerical differentiation in action, to see how it can help you to understand the concept of differentiation and to find out about the dangers and difficulties inherent in the process.

The spreadsheet is often thought of as a tool just for handling money, but it is actually very good for illustrating and implementing a variety of general numerical methods, although it has some shortcomings for specialist use. You will find it useful to learn basic spreadsheet commands such as how to link cells together by formulae and how to draw graphs. Get in the habit of labelling carefully what it is you are doing on the sheet - it is easy to get lost when the links between cells become complicated.

DO THIS NOW! 7.4   A Spreadsheet activity

x	y(x)	Num Diff	h = 0.1 Set h here.
0	0	0.1	
0.1	0.01	0.3	Formula in C2 is =(B3-B2)/$E$1, then copied down column.
0.2	0.04	0.5	
0.3	0.09	0.7	
0.4	0.16	0.9	
0.5	0.25	1.1	Formula in B2 is =A2^2, then copied down column
0.6	0.36	1.3	
0.7	0.49	1.5	
0.8	0.64	1.7	
0.9	0.81	1.9	
1	1	2.1	
1.1	1.21	2.3	
1.2	1.44	2.5	
1.3	1.69	2.7	
1.4	1.96	2.9	
1.5	2.25		

Formula in A3 is =A2+$E$1, then copied down column

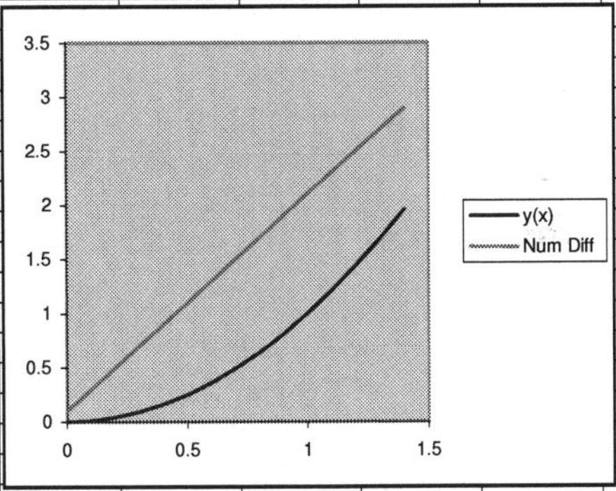

Figure 7.6 A spreadsheet to illustrate numerical differentiation.

To see numerical differentiation in action, look at Figure 7.6. This particular diagram was produced on Excel. Note the usual adding of $ signs to fix cell references.

Column A contains x values, set up to change as step length h is changed. Column B contains the function definition, in this case $y = x^2$ again. Column C contains the numerical derivative $\dfrac{y(x+h) - y(h)}{h}$. Both y and its numerical derivative are plotted.

Set up a sheet like this for yourself, vary h, and observe how:

(a) the slope is more accurately calculated as h becomes smaller;

(b) *if h gets too large then the derivative values calculated are very unreliable.*

Finally change the function being studied in column B to another of your choice and again experiment with changing h. Predict what shape you think the derived function curve should be, plot it, and explain why it agrees or not with your prediction.

7.6 PAPER AND PENCIL APPROACHES TO DIFFERENTIATION

There are rules for differentiating functions algebraically, which you can look at and gain some practice with in this section. We shall actually derive only one rule

algebraically, that for differentiating $y = x^2$, because you have already explored this function numerically and have some idea of how it behaves.

We shall then simply present a table of the other common standard derivatives without proof.

7.6.1 The rule for differentiating $y = x^2$

The derivative of $y = x^2$ is given by

$$\frac{dy}{dx} = \underset{h\to 0}{Lim}\left(\frac{y(x+h) - y(x)}{h}\right) = \underset{h\to 0}{Lim}\left(\frac{(x+h)^2 - x^2}{h}\right) = \underset{h\to 0}{Lim}\left(\frac{x^2 + 2xh + h^2 - x^2}{h}\right)$$

$$= \underset{h\to 0}{Lim}\left(\frac{2xh + h^2}{h}\right) = \underset{h\to 0}{Lim}\left(\frac{h(2x+h)}{h}\right) = \underset{h\to 0}{Lim}\left((2x+h)\right) = 2x$$

If you look at what is happening, you cannot simply put $h = 0$ at the beginning, because that would result in 0 divided by 0 and the answer is not determinable. You have to simplify the expression inside the brackets until it is clear how the various h terms interact. When the h in the denominator cancels out, then it is possible to put the value of 0 in for h, as you are no longer dividing by that.

7.6.2 Rules for other functions

You may derive for yourself some of the other standard derivatives in the table in Figure 7.7, using limiting processes as above. It is usual however just to use such tables to look up the standard derivatives. Note all derivatives are presented "with respect to x". This just means that x is the independent variable. If a different independent variable is used then just replace x. Thus for example, if $y = x^3$ then $\frac{dy}{dx} = 3x^2$. So also if $y = t^3$ then $\frac{dy}{dt} = 3t^2$, and so on.

$y(x)$	$\dfrac{dy}{dx}$
constant k	0
kx	k
x^n	nx^{n-1}
$e^{kx} = \exp(kx)$	$ke^{kx} = k.\exp(kx)$
$\ln(ax)$	$\frac{1}{x}$
$\sin(\omega x)$	$\omega\cos(\omega x)$
$\cos(\omega x)$	$-\omega\sin(\omega x)$

Figure 7.7 Table of standard derivatives (k, a, n, and ω are all constants)

Here are some examples of using the tables to differentiate some basic functions. Note that there are several different notations used to represent differentiation. You must get familiar with all of them. Note the useful shorthand y' for $\frac{dy}{dx}$

Example 7.6.1 $\dfrac{d}{dx}(e^x) = e^x$ $[\dfrac{d}{dx}()$ is so called "operator" notation and means "differentiate with respect to x whatever follows in the brackets"]

Example 7.6.2 If $y = \sin(2x)$, then from the tables with $\omega = 2$, $y' = \dfrac{dy}{dx} = 2\cos(2x)$.

Now here are some to practice on.

DO THIS NOW! 7.5
Use the tables to carry out the following differentiations, checking your own answers either using a CAS, or graphically.

1. $y = x^5$. Find $\dfrac{dy}{dx}$.

2. $y = \sqrt{x}$. Find $\dfrac{dy}{dx}$.

3. Find $\dfrac{d}{dx}\left(\dfrac{1}{x^3}\right)$.

4. $x = \cos(3t)$. Find $\dfrac{dx}{dt}$.

5. Find $\dfrac{d}{dx}(\ln(2.5x))$.

6. $y = e^{-2x}$. Find y'.

7.6.3 Rules for differentiating various combinations of functions

There are four rules for dealing with combinations of standard functions.

The sum and constant rule

$$\boxed{\dfrac{d}{dx}(au + bv) = a\dfrac{du}{dx} + b\dfrac{dv}{dx}}$$ where u and v are functions of x, and a and b are constants.

Example 7.6.3   Find $\dfrac{d}{dx}(2e^x + 4\sin x)$.

Note that $a = 2$ and $b = 4$, and $u = e^x$ and $v = \sin x$, and so:

$$\dfrac{d}{dx}(2e^x + 4\sin x) = 2e^x + 4\cos x$$.

The product rule. Spot the "×"

This rule allows you to find the derivative of a product of two functions. It is

$$\boxed{\dfrac{d}{dx}(u.v) = u.\dfrac{dv}{dx} + v.\dfrac{du}{dx}}$$,   where u and v are functions of x.

Example 7.6.4　　　　　　Find $\dfrac{dy}{dx}$ if $y = x.\sin(x)$.

First identify this as requiring the product rule. It can be helpful to tabulate the parts of the formula; it does not matter which way round you choose u and v, since $uv = vu$. It is critical to notice that although the multiplication sign may not be written, it is implied.

Choose $u = x$.	Choose $v = \sin(x)$.
Then $\dfrac{du}{dx} = 1$.	Then $\dfrac{dv}{dx} = \cos(x)$.

Thus $\quad \dfrac{dy}{dx} = \dfrac{d}{dx}(u.v) = u.\dfrac{dv}{dx} + v.\dfrac{du}{dx} = x.\cos(x) + \sin(x).1 = x\cos(x) + \sin(x)$.

The quotient rule.　Spot the " ÷ "

This rule allows you to find the derivative of a quotient of two functions. It is

$$\dfrac{d}{dx}\left(\dfrac{u}{v}\right) = \dfrac{v.\dfrac{du}{dx} - u.\dfrac{dv}{dx}}{v^2} \qquad \text{where } u \text{ and } v \text{ are functions of } x.$$

Remember to take the denominator v first and note the negative sign.

Often this rule is ignored since it is really a product of $u.v^{-1}$.

Example7.6.5　　　　　　Find $\dfrac{dy}{dx}$ if $y = \dfrac{\sin(x)}{x}$.

First identify this as requiring the quotient rule. Once more a tabulation can be helpful, but this time it matters which is u and which is v:

Choose $u = \sin(x)$.	Choose $v = x$.
Then $\dfrac{du}{dx} = \cos(x)$.	Then $\dfrac{dv}{dx} = 1$.

And so applying the rule: $\quad \dfrac{dy}{dx} = \dfrac{d}{dx}\left(\dfrac{u}{v}\right) = \dfrac{v.\dfrac{du}{dx} - u.\dfrac{dv}{dx}}{v^2} = \dfrac{x.\cos(x) - \sin(x).1}{x^2}$.

The chain rule or function of a function Rule　　Spot the "[( )]"

This allows you to find the derivative of a function which is built up from two functions, one "inside" the other. People often find this rule a little harder to use.

It is $\qquad \dfrac{dy}{dx} = \dfrac{d}{dx}[F(u(x)] = \dfrac{dF}{du}.\dfrac{du}{dx}$

Interpret what this means: differentiate the "outside" function as if it had just an x inside it. Then, to allow for the fact that it is not just an x, multiply by what you get if you differentiate the "inside" function.

Example 7.6.6

Find $\dfrac{dy}{dx}$, if $y = e^{x^2} = \exp(x^2)$. First, identify this as requiring the chain rule - this can be the hardest part! Once this is done, again a tabulation helps:

Inside function	Outside function
$u = x^2$	$F(u) = \exp(u)$
$\dfrac{du}{dx} = 2x$	$\dfrac{dF}{du} = \exp(u) = \exp(x^2)$

Finally then, just multiply the two bottom lines together:

$$\frac{dy}{dx} = \frac{d}{dx}[F(u(x))] = \frac{dF}{du}.\frac{du}{dx} = \exp(x^2).2x = 2xe^{x^2}$$

These are the rules which, together with your favourite set of tables of derivatives, should enable you to do any messy combination of functions. Note that with CAS technology around (section 7.7) you should always get these derivatives **correct**. Sometimes, answers in books may not look like yours, since there are many different forms of the same answer. One test is to graph both answers and, from the result, you should be able to "see" that you are correct. If there is no agreement, then checking, possibly algebraic manipulation, and discussion with others is necessary.

Example 7.6.7

If $y = \dfrac{\sin(x)}{x^2}$, with $x \neq 0$ (why is this needed?), straightforward use of quotient rule

gives $\dfrac{dy}{dx} = \dfrac{x^2 \cos(x) - \sin(x).2x}{(x^2)^2}$ but TI-Interactive! gives $\dfrac{dy}{dx} = \dfrac{\cos(x)}{x^2} - \dfrac{2\sin(x)}{x^3}$.

These look different, but are equivalent - explore this issue now.

DO THIS NOW! 7.6

In example 7.6.7, use your technology to plot both versions of the derivative against x, and convince yourself they are the same.

In example 7.6.7, show algebraically that the two answers are equivalent, by starting with the first form and using algebraic steps to reduce it to the second form. Make sure you explain and can verify each step you take.

In example 7.6.7, check and explain to yourself how the shape of the graph of $\dfrac{dy}{dx}$

against x (either form, since they are the same!) does indeed represent the slope of the graph of y against x. (Hint: plot both graphs on the same axes.)

DO THIS NOW! 7.7

Here are some general differentiations to practice on. Differentiate with respect to the appropriate variable and verify your own answers by other means (graphical perhaps, or using a Computer Algebra System or CAS).

(a) $y = 2\sin(4x)$

(b) $y = 2x^3 + 5x^2 - 3x + 6$

(c) $y = \dfrac{2}{x} - \dfrac{5}{x^2}$

(d) $y = \sqrt{x} + \dfrac{1}{\sqrt{x}}$

(e) $y = t.e^{-2t}$

(f) $y = e^{-t}\sin(t)$

(g) $y = \dfrac{\cos(\omega)}{\omega^2}$

(h) $y = \sin(x^2 + 1)$.

7.7 COMPUTER ALGEBRA SYSTEMS OR SYMBOL MANIPULATORS

The Computer Algebra Systems (CAS) is now commonplace. You should become comfortable with it, as well as developing your pencil and paper methods, and becoming comfortable with handling symbols. You can use CAS to check calculations, and use your calculations to reinforce and understand CAS output.

There are two general points here. One, is that you should not mindlessly accept what any technology tells you, but always find an independent way of verifying. The second is crucial and concerns what you put in. If you cannot communicate correctly with a CAS, the results you get out will not be reliable.

Consider for example, the problem of finding $\dfrac{d}{dx}(\sin^2(x))$. If you feed into CAS the function sin^2(x), most of them will not "understand" you. You would have to express this as (sin(x))^2. It seems, then, that the quirks of computer-based systems are sharpening up the notation we use. Note the shaded part of Figure 7.8 which illustrates how careful you have to be in communicating with technology (this is from a graphic calculator with CAS). When in doubt, use brackets!

Figure 7.8 Take care when entering expressions into CAS.

DO THIS NOW! 7.8

We suggest that you learn enough about your local CAS to check **NOW** the enhanced table of derivatives in Figure 7.9. You can also add other functions of your own choice and go on adding as you need to in the future. Use the techniques in **DO THIS NOW 7.6!** if your CAS answer differs from that in the table.

For a sample of these, plot both the original and the derived functions, and satisfy yourself that the derived function really does give the slope of the original function.

Table of standard derivatives

(Note k, a, n, and ω are all constants in this table.)

$y(x)$	$\dfrac{dy}{dx}$
constant k	0
kx	k
x^n	nx^{n-1}
$e^{kx} = \exp(kx)$	$ke^{kx} = k.\exp(kx)$
$\ln(ax)$	$\dfrac{1}{x}$
$\sin(\omega x)$	$\omega\cos(\omega x)$
$\cos(\omega x)$	$-\omega\sin(\omega x)$
$\tan(\omega x)$	$\omega\sec^2(\omega x) = \dfrac{\omega}{\cos^2(\omega x)}$
$e^{-kx}\sin(\omega x)$	$\omega e^{-kx}\cos(\omega x) - ke^{-kx}\sin(\omega x)$
$e^{-kx}\cos(\omega x)$	$-\omega e^{-kx}\sin(\omega x) - ke^{-kx}\cos(\omega x)$
$\dfrac{\sin x}{x}$	$\dfrac{x.\cos(x) - \sin(x).1}{x^2}$
e^{x^2}	$2xe^{x^2}$

Figure 7.9 Table of Standard Derivatives.

8

Differential Calculus 2

I'm very good at integral and differential calculus, I know the scientific names of beings animalculous; In short, in matters vegetable, animal, and mineral, I am the very model of a modern Major-General. Gilbert, W. S.

8.1 DIFFERENTIAL CALCULUS: TAKING THE IDEAS FURTHER

In this chapter you will get the chance to explore further the ideas you met in Chapter 7 on differential calculus. Amongst other activities you will look at the problems which were introduced at the beginning of that chapter and see how the ideas of calculus can be used to solve them; this will include looking at simple optimizing techniques. You will also meet a variety of extensions to ways of thinking about differentiation. This will include higher derivatives, parametric differentiation, implicit differentiation, and partial differentiation. Finally you will have the chance to work through a rather more extended case study.

We begin then by looking again at the short practical examples introduced at the start of Chapter 7. Once more remember to **do all the exercises and activities**.

8.2 SOLVING THE EXAMPLES FROM CHAPTER 7

Example 8.2.1: Solving Example 7.2.1 - the Maxbox problem

The first study is a classical problem known commonly as the maxbox problem. It may appear to be somewhat artificial but you may know that the flat-pack cardboard box industry (making pizza boxes and so on) is highly successful, and economic use of resources is an important aspect of that success. In this problem of optimizing volume with a given area of material you should note a range of ways to deal with the situation - by hand, using a graphic calculator, using spreadsheet, using a CAS, with or without the need for calculus.

Often the hardest part is to formulate the problem to express it mathematically – this is just as much a part of mathematics as handling the algebra. Using mathematics to describe engineering processes is really what you are aiming to be able to do. Take note that we are using pictures to help to understand the situations as much as possible, and the technology can often help here.

Recall the problem was this: an A4 sized piece of card (29.7cm by 21cm) has four identical square pieces with side of length x cm removed from the corners so that it forms a template for an open-topped box. Find the dimensions of the box with the largest volume. This situation is illustrated in Figure 8.1.

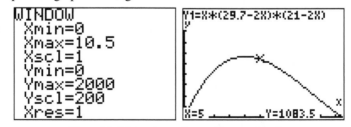

Figure 8.1 The maxbox problem

In formulating the problem mathematically, choice of variable is important, and we decide to say that the removed squares have sides of length x – this is because once this is decided then you can express the volume $V(x)$ in cm^3 by

$$V(x) = x(29.7 - 2x)(21 - 2x) \ .$$

The maximum possible value of this volume can be found in various ways and this is where we can explore the diversity of approaches.

a) Solving the Maxbox problem using a graphic calculator

We use the graphic calculator to plot $V(x)$ against x and look for the maximum. Take a moment to think about the sensible range over which to plot $V(x)$. x cannot be less than 0; nor can it be bigger than 10.5 cm (why not?). Thus we set the plotting range and plot the graph as in Figure 8.2:

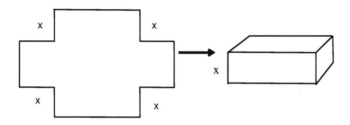

Figure 8.2 Solving the maxbox problem graphically.

Now you could use any built-in trace and zoom facility to identify the maximum volume in this range, or perhaps your calculator has a built-in maximum finder (Figure 8.3):

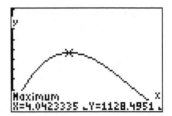

Figure 8.3 The maximum finder on a TI-83 Plus.

The answer emerges that the size of square to be removed is $x = 4.04$ cm, and that the maximum volume corresponding to this is $V(x) = 1128.5$ cm^3.

b) The Maxbox problem classical solution, algebraically on paper

You may note from Figure 8.3 that at the so-called "local" maximum point, the function $V(x)$ is horizontal – that is, it has a zero slope. We could therefore identify the point by looking for places where $\dfrac{dV}{dx} = 0$, as in this series of simple steps.

Multiply out the expression for $V(x)$ giving: $V(x) = 4x^3 - 101.4x^2 + 623.7x$.

Differentiate the result: $\qquad\qquad\qquad\qquad \dfrac{dV}{dx} = 12x^2 - 202.8x + 623.7$.

Equate this derivative to zero and solve the resulting quadratic equation for x:

$$x = 4.04 \text{ and } x = 12.86 \quad .$$

Note that $V(x)$ is a cubic and so the overall cubic shape gives Figure 8.4:-

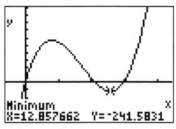

Figure 8.4 The function $V(x)$ plotted over a wider range.

Only one of these solutions, $x = 4.04$, lies within the allowable range $0 \le x \le 10.5$ and so given what we know about the shape of the curve it must give the maximum. The corresponding value of volume is $V(4.04) = 1128.5$. This agrees with what we found in (a).

For a closer look at finding and identifying maximum and minimum points on a curve see section 8.4.

c) Using a symbolic manipulation package to solve the Maxbox problem

Having formulated the problem, a CAS system can be used to carry out the algebraic details, as for example in Figure 8.5. This shows the volume function defined, the point of zero slope identified, and the calculation and use of the second derivative, which is discussed in sections 8.3 and 8.4.

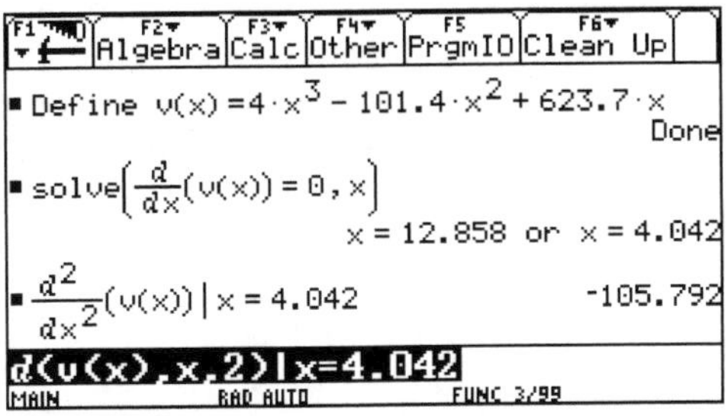

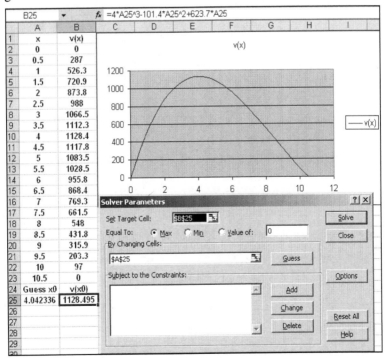

Figure 8.5 Using CAS as part of solving the maxbox problem.

DO THIS NOW! 8.1
Use your CAS system to carry out and check the algebraic steps in producing the answer to the Maxbox problem.

---

d) Using a spreadsheet to solve the Maxbox problem

A spreadsheet can be used to plot the function, Excel has a Solver facility on the Tools menu which you can use to find a maximum or minimum, as shown in Figure 8.6. It is useful to plot the function first, as for successful use the solver needs a good initial guess at the maximum.

Figure 8.6 Using the Excel Solver to find the maximum.

Example 8.2.2: Solving Example 7.2.2 - another optimisation problem

Example 7.2.2 is another variation on the theme of optimisation. An emergency petrol tank is designed to carry 1 gallon of petrol (4546cm^3). Its shape can be considered a cuboid. The base of the cuboid is a rectangle with the length double the width. Find the dimensions of the tank that minimise the surface area required.

Once more the difficult part is to formulate it. Take it one step at a time:

The first thing is to name the variables. Let the tank be x cm. wide by y cm. long on the base, by z cm. high, and let the volume be V cm^3.

Then $$V = xyz = 4546 . \tag{8.1}$$

If the base is x by y, then $y = 2x$. (Note x and y are interchangeable!) (8.2)

The surface area A = sum of four vertical sides + top + bottom,

that is, $$A = 2(xz + yz + xy) . \tag{8.3}$$

This looks messy until you realise that you can use (8.2) to remove y from the problem.

Thus (8.1) becomes $$V = 2x^2z = 4546 . \tag{8.4}$$

and (8.3) becomes $$A = 2(xz + 2xz + 2x^2) = 2(3xz + 2x^2) . \tag{8.5}$$

Following on, you realise that you can use (8.4) to remove z now:

(8.4) gives $$z = \frac{4546}{2x^2} .$$

and so from (8.5) $$A = 2(3x.\frac{4546}{2x^2} + 2x^2) = \frac{13638}{x} + 4x^2 .$$

This A is now the function for which we need to find the local minimum.

DO THIS NOW! 8.2

Finish off this problem to find the value of x giving minimum surface area A and hence work backwards to find y and z. Use at least one of the techniques in Example 8.2.1, and preferably more than one, to check. (We got $x = 11.95$ cm. Do you agree?)

The next examples show that often to "do maths" you need to read the words.

Example 8.2.3 Solving Example 7.2.3 - a suspension system model

A suspension system has displacement y given at time t by the function

$$y = e^{-0.5t} \sin(8.5t)$$

What is its speed (rate of change of position or displacement) at time $t = 6$ seconds?

It is apparent that the rate of change of displacement y will give the speed, so what we need here is the value of $\frac{dy}{dt}$ when $t = 6$ seconds. Using the product rule:

$$\frac{dy}{dt} = -0.5e^{-0.5t}\sin(8.5t) + 8.5e^{-0.5t}\cos(8.5t) \text{ and when } t = 6,\ \frac{dy}{dt} = 0.297 \ .$$

DO THIS NOW! 8.3

Check our answer by hand and by CAS (as e.g. in Figure 8.7). Do you get the same answer? If not, have you checked your calculator mode - are you working in radians as you should be?

```
■ d/dt(e^(-.5·t)·sin(8.5·t))
   8.500·(.607)^t·cos(8.500·t) - .500·(.607)▸
■ ◂- .5·(.60653065971264)^t·sin(8.5·t) | t = 6
                                            .297
```

Figure 8.7 Check your algebra using CAS.

Example 8.2.4 Solving Example 7.2.4 - an electronic problem

In simple electronics you may have come across the relationship $I = \frac{Q}{t}$, where I is the current, Q is charge and t is time. This is only true for constant current. If the current varies with time then the relationship is $i(t) = \frac{dq}{dt}$ = rate of change of charge with respect to time.

In one circuit, the charge varies according to $q(t) = 0.1e^{-0.1t}\sin(5t)$. What is the maximum value reached by the current?

DO THIS NOW! 8.4

In Example 8.2.4, find the current by differentiating the expression for $q(t)$. Using one of the techniques you have met, find its maximum value. Check it by another method. (We got a maximum value $y = 0.097$ when $t = 0.31$. Do you agree?)

8.3 HIGHER ORDER DERIVATIVES AND THEIR MEANING

Once you can differentiate a function then you can generally go on and differentiate the result again. For a function $y(x)$, the first derivative with respect to x is written $\frac{dy}{dx}$. If you now differentiate this result you obtain the so-called *second derivative* $\frac{d}{dx}\left(\frac{dy}{dx}\right) = \frac{d^2y}{dx^2}$. Note the notation and how it arises.

You can repeat this process to get the third derivative $\frac{d}{dx}\left(\frac{d^2y}{dx^2}\right) = \frac{d^3y}{dx^3}$, etc.

This raises the matter of what meaning to give to what you are doing. The first derivative gives the rate of change of a function, or the slope of its graph. If you differentiate again, then you are getting the rate of change of the rate of change, or how quickly the slope is changing as you move in the x-direction!

Example 8.3.1

Note that in the highly profitable industries of rollercoaster and lift design, in (position s) versus (time t) diagrams, these higher derivatives are useful.

Familiar quantities are velocity (rate of change of displacement) $v = \dfrac{ds}{dt}$, and

acceleration (rate of change of velocity) $a = \dfrac{dv}{dt} = \dfrac{d^2s}{dt^2}$. Less familiar quantities are

jerk (rate of change of acceleration) $j = \dfrac{da}{dt} = \dfrac{d^2v}{dt^2} = \dfrac{d^3s}{dt^3}$, and perhaps strangest of

all is jounce (acceleration of acceleration) $j_0 = \dfrac{dj}{dt} = \dfrac{d^2a}{dt^2} = \dfrac{d^3v}{dt^3} = \dfrac{d^4s}{dt^4}$.

The names are suggestive of their significance and of course requirements for their behaviour are quite different in the two industries!

Geometrically, the second derivative can be given meaning by talking about how quickly the slope of a curve is changing as you move in the x-direction. $\dfrac{d^2y}{dx^2}$ is

clearly then going to be related to how "curved" a function is. If $\dfrac{d^2y}{dx^2}$ is large in

magnitude then the slope is changing quickly and the function is very "curved". If it is small in magnitude then the function is not very curved.

Example 8.3.2

For a straight line, $y = mx + c$, the slope is $\dfrac{dy}{dx} = m$ and the second derivative

$\dfrac{d^2y}{dx^2} = 0$. This value of 0 for the rate of change of slope makes sense as a straight

line is not very curved - the slope does not change!

For a quadratic function $y = x^2$, the slope is $\dfrac{dy}{dx} = 2x$ and the second derivative

$\dfrac{d^2y}{dx^2} = 2$. In this case the rate of change of slope is 2. That is, the slope doubles for

every unit distance moved in the x direction. This accords with your knowledge of the quadratic curve, where the slope is large and negative to the left, but steadily increases (i.e. becomes more positive) through zero to become progressively more and more positive as you move to the right.

DO THIS NOW! 8.5

For the function $y = x^3 + x^2 - x - 0.5$, find $\dfrac{dy}{dx}$ and $\dfrac{d^2y}{dx^2}$ by hand. Check by another means.

Study the graphic calculator output of Figure 8.8. Note that Y_1 is y, Y_2 is $\dfrac{dy}{dx}$ and Y_3

is $\dfrac{d^2y}{dx^2}$. Work out which curve is which on the graph by studying the relationships

between the graphs. Check by plotting them yourself.

 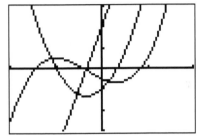

Figure 8.8 Match the curve to the derivative.

DO THIS NOW! 8.6

The equation of motion of one of a set of automatic doors in a shop is modelled, in terms of its displacement y from its fully closed position at time t, by the equation:

$$y = -41.3t^4 + 164t^3 - 213.7t^2 + 151t \qquad 0 \le t \le 2$$

By differentiating on paper and/or a CAS, find expressions for the velocity v, acceleration a, jerk j and jounce j_0 of the door's motion at time t. Verify that the maximum displacement occurs when $t \approx 1.84$ and that the jounce has a constant value of -991.2. By plotting graphs of y, v, a, j and j_0 against t, describe in words the motion of the door.

8.4 FINDING MAXIMUM AND MINIMUM POINTS

The process of finding maximum and minimum points needs a closer look. Such points where the slope is horizontal and the first deriviative is 0 are often called *turning* points or perhaps *stationary* points (from their meaning when applied in distance-time type problems). They are classified as *local* maxima (minima) if they are the biggest (smallest) value in the region, and *global* or overall maxima (minima) if they are the biggest (smallest) value the function takes anywhere. This classification can often be done either visually from the graph or by appealing to the real situation. For instance, in Example 8.2.1, the turning point at $x = 4.04$ is obviously a local maximum, both from the graph (for instance in Figure 8.2) and because of the physical nature of the situation. It is also however possible to deal with this identification mathematically.

There are two tests, the First Derivative Test and the Second Derivative Test.

8.4.1 The First Derivative Test

Evaluate the first derivative, or slope, at points just either side of the turning point. If the sign of the slope is positive to the left and negative to the right of the turning point then it is a maximum. This is illustrated in Figure 8.10 relative to Example 8.2.1. If the signs are negative to the left and positive to the right then it is a minimum. If the sign does not change then further graphical investigation is the best approach.

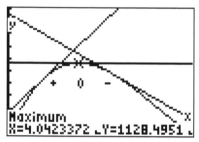

Figure 8.10 Identifying a maximum using the First Derivative Test.

8.4.2 The Second Derivative Test

Evaluate the second derivative at the turning point. If it is negative, it corresponds to a slope which is decreasing as you go from left to right - in other words a maximum. If it is positive, it corresponds to a slope which is increasing as you go from left to right - in other words a minimum. If it actually equals 0, then the test is inconclusive and graphical investigation is the best way forward.

DO THIS NOW! 8.7

For the function $V(x) = x(29.7 - 2x)(21 - 2x)$ from Example 8.2.1, perform both the first and second derivative tests to show that there is a maximum at $x = 4.04$.

Also perform both the first and second derivative tests to show that the turning point at $x = 12.86$ is a minimum.

8.5 PARAMETRIC DIFFERENTIATION

There are other ways to represent relationships apart from $y = f(x)$ – the so-called *explicit* function representation. One way is called *parametric representation*, which can be very useful in writing down a mathematical representation of two-dimensional motion. In this case we still need to be able to calculate rates of change.

Example 8.5.1

A point moves around on the screen of a cathode ray oscilloscope, painting out a curve as it goes. Its position on the screen is controlled by two sets of electrical plates: one controls left/right position x as a function of time t, the other controls up/down position y as a function of time. Thus the curve drawn by the point (x, y) as it moves is governed by two functions $x(t)$ and $y(t)$. This is a parametric definition of the curve, with t being the parameter which controls the movement.

For instance if $x(t) = 2\sin(t)$ and $y(t) = \sin(4t)$, then Figure 8.12 shows the curve which traced out as t increases from 0 to 2π.

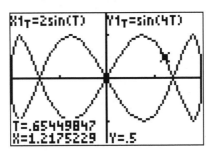

Figure 8.12 A parametrically defined curve.

DO THIS NOW! 8.8

Using your favourite technology, find out how it plots parametrically defined curves
NOW, and check that you understand by seeing if you can reproduce Figure 8.12.

Now given that a curve has been traced out, and that we know $x(t)$ and $y(t)$, what can
we say about the rate of change or slope of the x-y curve?

We could for instance find $\dfrac{dx}{dt} = 2\cos(t)$. This is the rate of change of x with respect

to t, or in other words, the speed of the point in the x direction as it traces out the

curve; similarly we could find $\dfrac{dy}{dt} = 4\cos(4t)$, which is the speed of the point in the

y direction as the curve is traced out.

From these we can use the formula $\dfrac{dy}{dx} = \dfrac{dy/dt}{dx/dt} = \dfrac{4\cos(4t)}{2\cos(t)}$, which we present

without proof, to get a formula for the slope of the curve in terms of t.

DO THIS NOW! 8.9

A curve is defined by the set of parametric equations $x(t) = \cos(t)$ and $y(t) = \sin(t)$.
First use algebra to find a connection between y and x not involving t (hint: square
each equation and then add them together). Use your technology in parametric mode
to plot the curve generated. Describe the curve you get. (Note you will not see the
curve properly unless you arrange for your x- and y-axis scales to be equal, and
therefore non-distorting.)

Calculate $\dfrac{dx}{dt}, \dfrac{dy}{dt}$ and $\dfrac{dy}{dx}$ in terms of t.

You should see a circle of radius 1 with its centre at the origin. Work out and mark

which points on the circle correspond to $t = 0, \frac{\pi}{4}, \frac{\pi}{2}, \pi, \frac{3\pi}{2}, 2\pi$. (Remember

RADIANS!). Calculate the slope at each of these points using your answer for $\dfrac{dy}{dx}$,

and satisfy yourself that these make sense in terms of your diagram.

(Note you should get $\dfrac{dy}{dx} = -\dfrac{\cos(t)}{\sin(t)}$.)

8.6 IMPLICIT DIFFERENTIATION

One way of understanding the idea of a function is to see it as a connection or relationship between two quantities, often labelled x and y, which gives a unique value of y for each value of x over the range for which it is defined. Sometimes that relationship is ***explicit***, e.g. $y = \sin(x)$, that is, given an x value you can immediately use the formula to find y. Sometimes the relationship is defined by a formula in which the x and y values are mixed up together. This is called an ***implicit*** function.

Example 8.6.1

The relationship $x^3 + y = y^2 + 2$ is an implicit definition. Note the difficulty of evaluating y given a value of x.

Example 8.6.2

A common example is the implicit definition $x^2 + y^2 = 1$ of a circle of radius 1 centred on the origin. This arises from the use of Pythagoras' Theorem, as shown in Figure 8.13.

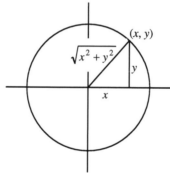

Figure 8.13 Defining a circle in terms of x and y.

It does raise the issue of the uniqueness of the y value for a given x value, because the formula can be rewritten $y = \pm\sqrt{1-x^2}$, and the square root can have either positive or negative values. Strictly speaking then, for this to be a *function* we would have to state that we are restricting ourselves to only positive or only negative values. To take account of both possibilities, mathematicians would use the word relation rather than function, as multiple values would then be allowed – the word function is used in a precise way.

Now having seen what an implicit definition looks like we can then ask how you find the slope of an implicitly defined function. The answer is that you simply use the rules of differentiation as developed in Chapter 7. Let us do it for the circle example.

Example 8.6.2 (continued)

The circle is defined by $x^2 + y^2 = 1$. To find the slope $\dfrac{dy}{dx}$, just differentiate the whole equation term by term with respect to x:

$$\frac{d}{dx}(x^2 + y^2) = \frac{d}{dx}(1) \ .$$

Rewriting $$\frac{d}{dx}(x^2) + \frac{d}{dx}(y^2) = \frac{d}{dx}(1) \ .$$

Differentiating $$2x + \frac{d}{dx}(y^2) = 0 \ .$$

(We are just using standard formulae or technology.)

We are left to differentiate a function of y, which itself is a function of x, with respect to x. You may recall the chain rule or function of a function rule (section 7.6), which says that you simply differentiate the "inside" function, then the "outside" function, and then multiply the results. In this case that can be interpreted in a simple way: that you differentiate the ("outside") y^2 term with respect to y to give $2y$; then differentiate the "inside" term y with respect to x to get $\frac{dy}{dx}$, then multiply them together. This gives

$$2x + 2y\frac{dy}{dx} = 0 \ .$$

Effectively you have treated the y term as if it were x, but have to remember to multiply by $\frac{dy}{dx}$ to take the y into account. You can see now that this can be solved for $\frac{dy}{dx}$ to give its value in terms of both x and y:

$$\frac{dy}{dx} = -\frac{x}{y} \ .$$

DO THIS NOW! 8.10
Plot the circle (remember you must have equal scales in x- and y-directions if your circle is to look undistorted), calculate some (x, y) points on it, and then substitute those values into the formula for $\frac{dy}{dx}$ to check that they make sense in terms of what the slope looks like. Compare it with your results from DO THIS NOW 8.9!

DO THIS NOW! 8.11
The magnetic bottle function – used to model containment of hot metals - is

$$x^{\frac{2}{3}} + y^{\frac{2}{3}} = 1 \ .$$

Find an expression for its slope.

Its parametric form is $x(t) = \sin^3(t), y = \cos^3(t)$. Use this form to sketch the curve, and then use the curve to check that your expression for the slope makes sense.

8.7 PARTIAL DIFFERENTIATION
8.7.1 Functions of more than one variable

Sometimes one quantity will depend on more than one other variable – in this case we say we have a function of more than one variable. This raises two questions: how can we build a concept of these functions in our minds; and what does rate of change mean in this context?

8.7.2 A geometrical way of thinking about functions of two variables

A good conceptual way of thinking about functions of two variables is to think in terms of x and y as position east/west and north/south on a flat map, and of z as being the height of the land at each point (x, y) on the map. One example is shown in Figure 8.14. The height z is then a function of two variables $z(x, y)$.

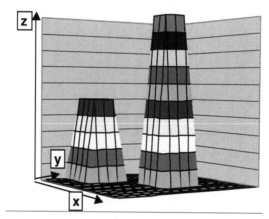

Figure 8.14 A way of visualising functions of two variables.

Example 8.7.1

Figure 8.15 shows a plot of the function of two variables $z = x^2 + y^2$.

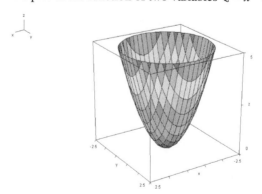

Figure 8.15 A Derive plot of a function of two variables.

Note that in this case $z(x, y)$ will always be non-negative.

Example 8.7.2

Figure 8.16 shows a plot of $z = \sqrt{1 - x - y}$.

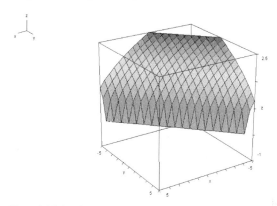

Figure 8.16 Another Derive plot of a function of two variables.

Once more note the range here – since z must be purely real to be plotted, we require that the expression inside the square root must not be negative. Thus we must have $1 - x - y \geq 0$.

DO THIS NOW! 8.12

Find out NOW which of your mathematical technologies will plot functions of two variables and find out how to do it. Test whether you understand by reproducing the plots in Examples 8.7.1 and 8.7.2.

Explore Example 8.7.1 by calculating z at few values of x and y and seeing how these fit in with the technologically produced graph.

(Harder) If in Example 8.7.1 we put $y = 0$, then the function reduces to $z = x^2$. This corresponds geometrically to taking a vertical slice through the diagram along the direction $y = 0$, producing a 2 dimensional graph corresponding to $z = x^2$. Sketch what this graph looks like, arguing both from the picture and from the equation $z = x^2$.

Plot (thinking about what is a suitable range) the functions:

(a) $z = \sin(x + y)$ (b) $z = \sqrt{1 - x^2 - y^2}$.

8.7.3 Rates of change and functions of two variables

What do we mean by the rate of change of a function of two variables, $z(x, y)$? It depends on what the rate of change is with respect to. If we have a situation where we treat y as constant and differentiate the function with respect to x, we are finding

the rate of change of z as x changes with y constant. This is written $\left(\dfrac{\partial z}{\partial x} \right)_y$, with

curly ∂ being used rather than straight d to remind us that we are holding something constant. It is known as the partial derivative of z with respect to x at constant y. This is often written just $\dfrac{\partial z}{\partial x}$ if it is obvious what is being held constant. Geometrically it means finding the slope of the surface $z = z(x, y)$ in the x direction. Correspondingly you can find $\left(\dfrac{\partial z}{\partial y}\right)_x$ or $\dfrac{\partial z}{\partial y}$, being the rate of change or the slope in the y direction.

Example 8.7.3

Return to $z = x^2 + y^2$, illustrated in Figure 8.15. In this case

$$\frac{\partial z}{\partial x} = 2x + 0 = 2x \ ,$$

and

$$\frac{\partial z}{\partial y} = 0 + 2y = 2y \ .$$

One immediate point is that these partial derivatives can depend on both x and y – all that is saying is that the slope of the surface $z = z(x, y)$ varies from one point to another, and that is obvious if you look at Figure 8.15.

It is useful to compare Figure 8.15 with the expression for the x direction slope $\dfrac{\partial z}{\partial x} = 2x$. You can say that when x is zero the slope in the x direction is zero: this means someone walking on the surface in the positive x direction would feel themselves neither climbing nor descending as they pass $x = 0$. If x is positive, then so is the slope, and as x gets bigger then so does the slope: this means someone walking in the positive x direction with $x > 0$ would experience climbing, with the slope getting steeper as you go further. Correspondingly, if x is negative then the slope is negative and our walker moving in the positive x direction would experience walking downhill, with the slope being steeper further back in negative x.

This idea is illustrated in Figure 8.17. If you take a section through a surface as shown, say along the plane at $y = 1$, you can see how the slope of the surface varies in that direction, and in fact you can plot the section as a graph in 2-dimensions.

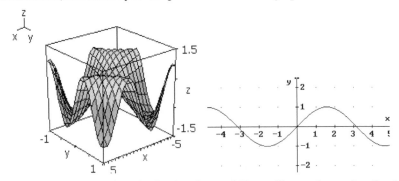

Figure 8.17 A cross section through a function of two variables too illustrate the meaning of partial derivative.

DO THIS NOW! 8.13

In each of the following cases, find the two partial derivatives $\dfrac{\partial z}{\partial x}$ and $\dfrac{\partial z}{\partial y}$

(a) $z = \sin(x + y)$ (b) $z = \sqrt{1 - x^2 - y^2}$.

You have already plotted these in DO THIS NOW 8.12! Describe using words and pictures how you see that the expressions you have found for the partial derivatives do correspond to the shape of the functions you have plotted.

8.8 AN ENGINEERING CASE STUDY

In this section you have an opportunity to work through a real case study.

DO THIS NOW! 8.14

Have an uplifting experience!

Some of the main aims of the position control system in a lift are: to bring the lift car to the correct floor: to minimise travel time; to maximise passenger comfort by providing a smooth ride; to accelerate, decelerate and travel within safe speed limits.

Acceptable functions for a "flight path" to describe vertical position $s(t)$ (metres) with respect to time t in seconds for $0 \le t \le 40$ are given by:

$$s_1(t) = -0.00009t^4 + 0.0094t^3 - 0.3703t^2 + 6.3723t$$

and $s_2(t) = 0.00008t^4 - 0.0078t^3 + 0.2167t^2 - 0.2741t$.

Using your knowledge of the meaning and use of differentiation, decide which is the preferable function, describing in your own words why.

Hint:- look at the acceleration and use the ideas of section 8.3. Bigger hint: we think the second one provides the best ride – do you agree?

END OF CHAPTER 8 EXERCISES – DO ALL THESE NOW!

Check your own answers by confirming answers both algebraically and graphically and, where you can, seeing if they are physically sensible.

1. Plot the function $y = x^3 + 2x^2 - 6x - 2$ and identify graphically and using calculus methods any local maximum and minimum points.

2. Find the number of revolutions per minute (n) which gives the minimum frictional couple (F) on a bearing where:
$$F = (180000 - 1200n - 45n^2 + n^3) / 80000 .$$

3. A clockwork toy train is released from. Its speed, V m s^{-1}, varies with time t and is given by
$$V = 0.17(t - 0.0007t^3) \quad (0 \le t \le t_{max}),$$
where t seconds is the time since the train was released, and t_{max} seconds is the time the train takes to come to a halt. What is the value of t_{max}? Find the time when the speed of the train will be at a maximum.

4. (Harder) Two corridors, 4m and 2m wide, meet at right angles. Find the length of the longest scaffold pole which can be carried round the corner in an horizontal position.

5. A wooden packing case is made in the form of a rectangular block without a lid and standing on a square base. Its volume is 0.2 m³. If x m is the length of a side of the base, find the value of x for which area of wood is a minimum.

6. If a ball is thrown up with a velocity of 70 ms⁻¹ the height x m reached in t s is given by $x = 70t - 4.9t^2$. Find the greatest height reached.

7. A cylindrical tank, closed at both ends, is to have the volume of 300cm³. Find its height and radius when the total surface area is a minimum.

8. There is an oil well 2 miles offshore and a refinery on shore at a point 5 miles down the coast from the closest point on the coast to the well. It costs £40000 per mile for pipe under water and £15000 per mile for pipe on the beach. At what point P should the pipe be brought to the coast to give the minimum cost? For what range of costs on land will it be best to take a direct path from the well to the refinery?

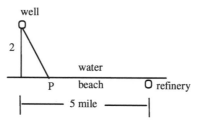

9. A wheel rolls with constant speed along a level road. The position of a point A on the rim at time t seconds is given by $x = 0.6(t - \sin t), y = 0.6(1 - \cos t)$ where x is the distance in metres measured horizontally and y is the distance in metres measured vertically upwards. When $t = 1$ second, find

   (a)   the horizontal velocity $\dfrac{dx}{dt}$ of the point A;

   (b)   the vertical velocity $\dfrac{dy}{dt}$ of the point A;

   (c)   the gradient $\dfrac{dy}{dx}$ of the curve along which A moves.

10. Find the first partial derivatives of each of the following:

   (i) $z = 2x^3 + 3xy + 4y^2$;

   (ii) $z = \cos(xy)$.

9

Integral Calculus 1

"There is too much preoccupation with what might be called the magic in calculus" Peter Lax

9.1 THE NEED FOR INTEGRAL CALCULUS

This section provides activities to introduce you to the ideas of the integral calculus and the process and concept of integration. There are two ways of viewing integration, and it is a remarkable thing is that these two views involve the same process.

- It is a summation process, and can be used to add up total areas, volumes, moments of inertia, centres of gravity, charges, stresses, strains and indeed any cumulative process where an overall effect is required but the quantities vary internally. To an engineer perhaps this is the most important property, but it is also necessary to know about the other aspect.

- It is the process of "undoing" differentiation - that is, given the derivative, what is the function it came from? Differentiation and integration are inverse processes - one undoes the effect of the other.

You will need to understand both views of integration, to interpret the ideas numerically and graphically, and to handle the process symbolically. The numerical and graphical idea concerns interpreting and understanding integration as a summation process. As with differentiation, the symbolic approach concerns a set of algebraic procedures which seem to have no relation to what the process means!

To gain some idea of what the important concept of integration involves, you should **work through all the activities and exercises in this chapter**. If you bring some previous experience of integral calculus with you, then this is an excellent chance to renew your knowledge and skills.

9.2 INDEFINITE INTEGRATION AND THE ARBITRARY CONSTANT

If two functions $f(x)$ and $g(x)$ are related by $\dfrac{dg}{dx} = f(x)$ then the inverse statement is written $g(x) = \int f(x)dx + C$ where C is an arbitrary constant.

That is, integration is the *inverse* of differentiation - if f is the derivative of g then g is the integral of f. Note the notation: the integral sign $\int (..)dx$ operates on the function $f(x)$. The dx is an essential part of the symbol and notation. It confirms what the integral is "with respect to" – that is it tells you what the independent variable is, and (in crude engineering terms) sets its units. The integral $\int f(x)dx$ is called the *indefinite* integral of $f(x)$, for reasons which will emerge shortly from **DO THIS NOW! 9.1**.

DO THIS NOW! 9.1

Perform the following three differentiations (using your technology or by hand):

(a) $\dfrac{d(\sin(x))}{dx}$ (b) $\dfrac{d(\sin(x)+1)}{dx}$

(c) $\dfrac{d(\sin(x)+C)}{dx}$ (C is any constant).

Which of these differentiations is reversed when you integrate $f(x) = \cos(x)$; that is, when you perform the integral $y = \int \cos(x)dx$?

You should have noted that all three of the functions when differentiated give the same result: $f(x) = \cos(x)$. Any added constants disappear completely when differentiated (the rate of change of a constant is zero!) so when you come to reverse the differentiation (to integrate), then you don't know if the constant was there or not. You must thus always allow that there might be a constant there, but you won't know its value unless you have extra information - that is, its value is arbitrary (not fixed), hence the name *arbitrary* constant.

If you do have extra information, such as $y = 2$ when $x = 0$, then you can identify the value of the constant, thus: $$y = \int \cos(x)dx = \sin(x) + C \ ,$$

and if $y = 2$ when $x = 0$ then $2 = \sin(0) + C$ and so $C = 2$.

These facts and ideas become important when you are dealing with differential equations (Chapter 15), and do make sense in engineering terms: a derivative tells you how a system varies or changes, but to get full information you need to know where the system started from too!

9.3 USING A COMPUTER ALGEBRA SYSTEM TO MAKE A TABLE OF INTEGRALS

You can build up a table of standard integrals just as you did with the table of derivatives. Actually, because integration is the inverse of differentiation, you could use the differentiation tables right to left, but it is more convenient to have separate integration tables. We shall look at a small set of paper and pencil integration techniques in Section 9.7, using the simpler functions in these tables as building blocks for more complicated functions, but for now get your table of integrals started using a CAS.

DO THIS NOW! 9.2

Use your own CAS to check and enhance the table of integrals in Figure 9.1. You can also add other functions of your own choice and go on adding as you need to in the future. Pick a sample of these and check they really do work by differentiating the result again and seeing if you get back the original function ($\pm$ an arbitrary constant, of course). Note your CAS may not show the arbitrary constant.

$f(x)$ $(a, b, k, n, A, \omega, \phi$ constants$)$	$\int f(x)dx$ $(C$ arbitrary throughout$)$
constant k	$kx + C$
kx	$\dfrac{kx^2}{2} + C$
kx^n $(n \neq -1$ - why?$)$	$\dfrac{kx^{n+1}}{n+1} + C$
$k(ax+b)^n$ $(n \neq -1$ - why?$)$	$\dfrac{1}{a}\dfrac{k(ax+b)^{n+1}}{n+1} + C$
$\dfrac{k}{x}$	$k\ln x + C$
$\dfrac{k}{ax+b}$	$\dfrac{k\ln(ax+b)}{a} + C$
ae^{-kx}	$\dfrac{ae^{-kx}}{-k} + C$
$A\sin(\omega x + \phi)$	$-\dfrac{A}{\omega}\cos(\omega x + \phi) + C$
$A\cos(\omega x + \phi)$	$\dfrac{A}{\omega}\sin(\omega x + \phi) + C$
$\dfrac{1}{(x+a)(x+b)}$	$\dfrac{1}{(b-a)}\ln\left(\dfrac{x+a}{x+b}\right) + C$

Figure 9.1 A table of integrals.

9.4 DEFINITE INTEGRATION AND AREAS.
9.4.1 Definite integrals and areas under curves

We now move to considering integration as a summation process. The process of integration calculates the area between the curve $y = f(x)$ and the x-axis between x values of a (the bottom limit) and b (the top limit), as shaded in Figure 9.2(a), limiting ourselves for now to $f(x) \geq 0$ between these limits. Why does this work?

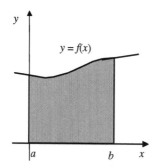
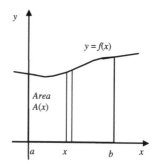

Figure 9.2(a) Finding the area.   Figure 9.2 (b) Illustrating the integration process.

To see why, it is best to imagine the area being built up by a vertical line varying in length as the height of the function, starting at $x = a$ and sweeping to the right across the area to $x = b$. This means the area will be built up as a function of x. Let this area under the curve between $x = a$ and any intermediate value x be called $A(x)$. Figure 9.2(b) shows what this looks like when the line has reached a general value x. It also shows that when the line sweeping out the area moves on by an additional small amount δx, then the area increases by a corresponding small amount δA. Note two things here. First, this splitting of the area into vertical strips will become a feature of our approach to understanding the use of integration. Next, it is usual, and you should get used to it, to use the Greek letter δ to denote a small change in a quantity. Returning to the diagram, the area of the additional strip is approximately the height y times the width δx of the strip, so the two small quantities are related by

$$\delta A \approx y \delta x$$

This may be rewritten

$$\frac{\delta A}{\delta x} \approx y$$

In the limit, as we let δx get smaller and smaller, the relationship becomes $\frac{dA}{dx} = y$. That is, height is the rate of change of area with respect to x. Turning this the other way round: $A = \int y \, dx + C$. That is, the area is the integral of the height of the curve above the x-axis. This says how area varies as you move along the curve.

It remains only to put in information about where the area starts and ends, and this works as follows.

The area is calculated using the so-called *definite integral* $\int_a^b f(x)dx$, which is defined in terms of the indefinite integral as follows:

$$\text{If } \int f(x)dx = g(x)+C \text{ then } \int_a^b f(x)dx = g(b)-g(a)$$

That is, do the indefinite integral, evaluate the result at the top and bottom limits, and take away the bottom result from the top result. Note that the limits appear at the top and bottom of the integral sign and the arbitrary constant disappears.

Example 9.4.1

Find the area between the curve $y = e^x$ and the *x*-axis between the values $x = 0$ and $x = 1$.

This area is given by $\text{Area} = \int_0^1 e^x dx = \left[\frac{1}{2}e^x\right]_0^1$.

We have done the indefinite integral, and note that we have carried the limits across from the integral – take note of the notation here. Now we complete the definite integral:

$$\text{Area} = \int_0^1 e^x dx = \left[e^x\right]_0^1 = e^1 - e^0 \approx 1.71828 .$$

Thus while the indefinite integral gives an algebraic expression, the definite integral is just a number.

9.4.2 Positive and negative areas

There is inevitably a complication, as you will find out now:

DO THIS NOW! 9.3

Plot the function $y = \cos(x)$ over the range $-\dfrac{\pi}{2} < x < \dfrac{3\pi}{2}$ (in radian mode!)

Evaluate the definite integrals (a) $\int_{\frac{\pi}{2}}^{\frac{\pi}{2}} \cos(x)dx$ and   (b) $\int_{\frac{\pi}{2}}^{\frac{3\pi}{2}} \cos(x)dx$.

Looking at your answers and your graph, delete as appropriate in the following:

Areas above the *x*-axis appear in an integral as ......(positive/negative).

Areas below the *x*-axis appear in an integral as ......(positive/negative).

You should see that an integral will calculate any area which lies below the *x*-axis as a negative quantity. The idea of a negative area may seem strange, but just remember that it simply means that it lies below the *x*-axis.

DO THIS NOW! 9.4

Predict the value of $\displaystyle\int_{\frac{\pi}{2}}^{\frac{3\pi}{2}}\cos(x)dx$ from your plot in DO THIS NOW 9.3! Explain your

prediction and verify it by doing the integral.

Finally, plot the function $y=x^2+3x+2$ over a suitable range, show by integrating

this function that the integral $\displaystyle\int_{-2}^{-0.5}x^2+3x+2\,dx$ has a value 0, and explain this result

in terms of areas above and below the x-axis.

So, in a case such as $\displaystyle\int_{\frac{\pi}{2}}^{\frac{3\pi}{2}}\cos(x)dx$, where it is clear from the curve plot that there

are equal amounts of area above and below the x-axis, the integral will predict a total
area of zero, as the positive area above the axis cancels out the negative area below.

9.4.3 Splitting the range of integration

Sometimes in engineering, systems behave differently at different times - they
must then be described by so called "piecewise defined functions".

Example 9.4.2

The graph of the function defined by $f(x) = \begin{cases} 1 & \text{for } x < 0 \\ x & \text{for } x \geq 0 \end{cases}$ is plotted in Figure 9.3.

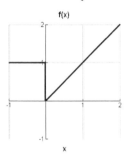

Figure 9.3 The graph of a piecewise defined function.

If we need to integrate this function, say between limits of -1 and 1, then we simply
split the range of integration up like this:

$$\int_{-1}^{1}f(x)dx = \int_{-1}^{0}f(x)dx + \int_{0}^{1}f(x)dx = \int_{-1}^{0}1dx + \int_{0}^{1}xdx = \left[x\right]_{-1}^{0} + \left[\frac{x^2}{2}\right]_{0}^{1} = 0-(-1)+\frac{1}{2}-\frac{0}{2}=1.5$$

This is OK since integration is just a summation process, so you just add up the
contributions to the area (integral) from the different sections of the curve – in this
case the areas are just a square and a triangle, so you can easily verify your answer..

DO THIS NOW! 9.5

You are given the function describing a half-wave rectified sine wave, defined by:

$$f(t) = \begin{cases} \sin(t) & 0 < t < \pi \\ 0 & \pi \leq t \leq 2\pi \end{cases}$$

with $f(t + 2\pi) = f(t)$ (this means the function repeats every 2π in the t direction.)

Plot the graph of $f(t)$ against t for $-2\pi \leq t \leq 2\pi$, and find the integral of $f(t)$ with respect to t between the limits of 0 and $\dfrac{3\pi}{2}$, splitting integration range as necessary.

The concept of a periodic function – one which repeats itself regularly – as introduced in **DO THIS NOW! 9.5** is a very useful one and you should note the notation used and ensure you understand its meaning.

9.5 USING AREAS TO ESTIMATE INTEGRALS.

If an integral can be evaluated analytically (that is, by algebraic methods), then its value (and the value of the area under the curve) can be found exactly. If it cannot, then you can do things the other way round - an estimate of the area will give an estimate of the value of the integral.

To illustrate this point, here is an example which can be done either by evaluating the integral or by finding the area.

Example 9.5.1

Evaluate $\displaystyle\int_0^3 2dx$, both by doing the integral, and by "seeing" the area.

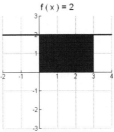

Figure 9.4 Finding the area both by integration and by seeing it.

By integral: from the tables the integral is $\displaystyle\int_0^3 2dx = [2x]_0^3 = 6 - 0 = 6$.

By area: it is evident from the picture that the integral value corresponds to the shaded rectangular area, which has a value $2 \times 3 = 6$.

We are delighted to find that the two methods agree about the size of the area!

DO THIS NOW! 9.6

Using your technology, and also by hand, evaluate the following definite integrals:

(i) $\int_{0}^{1} x\,dx$ (ii) $\int_{-1}^{0} x\,dx$ (iii) $\int_{-1}^{1} x\,dx$ (iv) $\int_{1}^{3} x\,dx$.

With reference to the graph of $y = x$, verify and justify your answers in terms of the areas of appropriate triangles and rectangles. Remember the area of a triangle is given by area = half (base×height)),

---

These examples so far are artificially easy, in that both the area and the integrals can be evaluated fairly easily. If $f(x)$ is more "curvy" then things are more complicated, in that the integral may be more difficult (or simply not possible), and the shape of the area may not have a simple formula associated with it.

What then is the best way to think about an integral as it calculates area "under the curve"? The brief discussion at the start of section 9.4 gives a clue. Developing the idea of a vertical line sweeping across the area, adding to the sum total of the area by one vertical strip at a time, you see that you can think of the process of integration as splitting the area into small strips of thickness δx and height $y = f(x)$ (i.e. lots of tall, thin, *almost* rectangles) for x values between a and b, and then adding up all the areas of these strips. This really emphasises the idea that integration is a summation process.

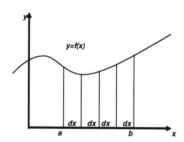

Figure 9.5 Conceiving of integration as "adding up the strips".

Figure 9.5 illustrates this idea and we go on to explore this a little more in the next section 9.6.

9.6 APPROXIMATE INTEGRATION - THE TRAPEZIUM RULE AND SIMPSON'S RULE.

You can think of an area "under the curve" as being divided into vertical strips of width h (this letter is often used when we are approximating, rather than δx), and the overall area is found by adding up the areas of the individual strips. Of course, the question then arises as to how to find the area of the each strip. This situation is represented in Figure 9.6.

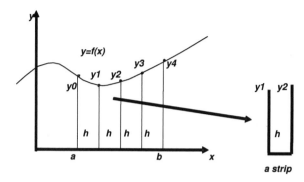

Figure 9.6 Evaluating area by adding areas of vertical strips.

The key to finding the area of each strip lies in what assumption is made about the shape of the top of each strip - it will in general be of different heights on each side. There are two commonly used rules. We present them with a brief explanation of their underlying assumptions, but not with a full derivation.

If the tops are joined with straight lines, you get the trapezium rule:

$$\text{Integral} \approx \frac{h}{2}(y_0 + 2y_1 + 2y_2 + 2y_3 + y_4) \qquad \text{trapezium rule.}$$

(This arises because each strip is a trapezium, with area equal to half the sum of the parallel sides times the distance between them, with the total area coming from adding up all of the strips.)

If the tops are joined with quadratics, you get Simpson's rule; (you must have an even number of strips here):

$$\text{Integral} \approx \frac{h}{3}(y_0 + 4y_1 + 2y_2 + 4y_3 + y_4) \qquad \text{Simpson's rule.}$$

(This formula is obtained by taking two strips at a time, fitting a quadratic curve through the top points and then integrating between the limits, giving $\frac{h}{3}(y_{left} + 4y_{middle} + y_{right})$, and then again adding up all the pairs of strips.)

Example 9.6.1

Find the area between the x-axis and the curve $y = e^{-x}$ over the range $0 \le x \le 1$ (Figure 9.7) using direct integration, the trapezium rule and Simpson's rule. Compare and discuss your answers.

Direct integration: Area $= \int_0^1 e^{-x} dx = \left[-e^{-x} \right]_0^1 = -e^{-1} - (-e^{-0}) = 1 - e^{-1} = 0.63212.$

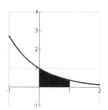

Figure 9.7 The area under an exponential.

If we are to use the trapezium rule or Simpson's rule, the first choice is the number of strips to use. We shall choose four strips (Figure 9.8) but explore this further shortly.

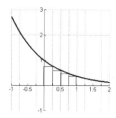

Figure 9.8 Numerical integration with four strips.

The value of h to use in the formulae is thus $h = 0.25$. The y values to use are presented in the following table:

x	0	0.25	0.5	0.75	1
y	$y_0 = e^0$ $= 1$	$y_1 = e^{-0.25}$ $= 0.779$	$y_2 = e^{-0.5}$ $= 0.607$	$y_3 = e^{-0.75}$ $= 0.472$	$y_4 = e^{-1}$ $= 0.368$

By trapezium rule: Area $\approx \dfrac{h}{2}(y_0 + 2y_1 + 2y_2 + 2y_3 + y_4) \approx 0.63541$.

By Simpson's rule: Area $\approx \dfrac{h}{3}(y_0 + 4y_1 + 2y_2 + 4y_3 + y_4) \approx 0.63213$.

You may like to verify these results by spreadsheet calculation, as a spreadsheet is ideally suited to regular numerical calculations such as this.

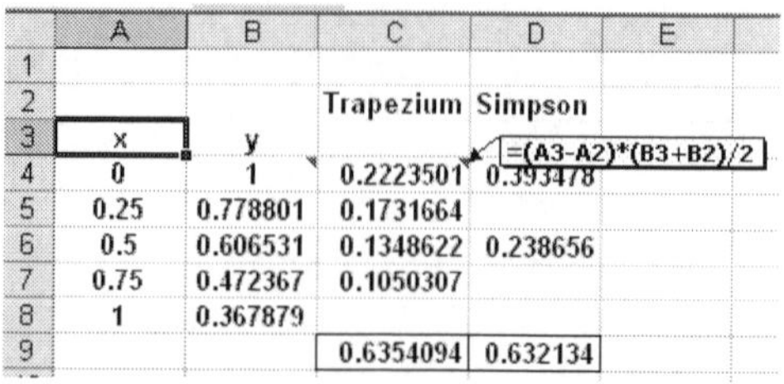

	A	B	C	D	E
1					
2			Trapezium	Simpson	
3	x	y			=(A3-A2)*(B3+B2)/2
4	0	1	0.2223501	0.393478	
5	0.25	0.778801	0.1731664		
6	0.5	0.606531	0.1348622	0.238656	
7	0.75	0.472367	0.1050307		
8	1	0.367879			
9			0.6354094	0.632134	

Figure 9.9 Numerical integration on a spreadsheet.

You see that the three values differ by a small amount, with the Simpson's rule estimate being closer to the actual value than that of the trapezium rule. This is because the quadratics used to create Simpson's rule can follow more closely than straight lines the actual function shape of the top of the strip as used in the integral.

However you might guess that the number of strips used also has an effect on the accuracy of these rules. This next exercise gives you the chance to explore this.

DO THIS NOW! 9.7

Plot $y = \cos x$ between the limits $0 < x < \dfrac{\pi}{2}$. Perform the integral $\displaystyle\int_{0}^{\frac{\pi}{2}} \cos x\, dx$

(a) analytically (by hand and/or technology);

(b) using trapezium rule with 2 strips;

(c) using trapezium rule with 4 strips;

(d) using Simpson's rule with 2 strips;

(e) using Simpson's rule with 4 strips;

(f) on a graphic calculator.

You may find it easier to do the numerical work on a spreadsheet. Compare and contrast the various estimates you get in the light of the facts, explaining any differences.

You should observe that

- Simpson's rule gives a more accurate estimate of the integral than the trapezium rule (as before);

- if more (and therefore narrower) strips are used, then the answer becomes more accurate.

This second point is important. If the strips are narrower then it is apparent that what you assume about the shape of the top becomes progressively less important – the closer you get to a curve the more it looks like a straight line anyway. See Figure 9.10.

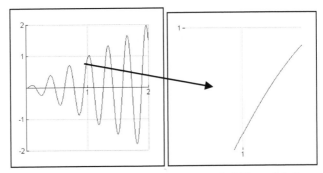

Figure 9.10 On a large enough scale, most curves look like straight lines.

In fact, if you allow the number of strips to get bigger and bigger then eventually the estimate provided by either rule can be made to get as close to the actual value of the integral as you wish. This is how the actual integral formula is theoretically constructed. A practical point is that, given the amount of computing power you have, then the number of strips can be increased as much as you like without much extra effort, so the trapezium rule is often quite adequate. When finding a definite integral you could for instance obtain an answer for a given step length and then halve the step length and compare the results. This process could be repeated to get the required accuracy. This is the approach programmed into graphical calculators.

9.7 PAPER AND PENCIL APPROACHES TO INTEGRATION

9.7.1 Some general remarks

Finally in this chapter, here is some work for you to do on paper and pencil approaches to integration. There are many different techniques for integrating by hand, and you will inevitably only meet a few here, but you should note that:

- many integrals occurring in engineering are definite, and are done numerically using perhaps the trapezium rule or something similar;

- indefinite integrals (as well as definite integrals) can be done using a CAS;

- not every integral can be evaluated analytically - this includes commonly useful integrals, for example $\int e^{-x^2} dx$ - and so these must be done numerically anyway.

Nevertheless it is worth getting some skills in integration to get a sense of how the variables behave and what answers mean.

9.7.2 Standard integrals from the tables

If you do want to do an integral by hand, the first hope is that it is in some tables - such as the one you started building up earlier.

Example 9.7.1

$$\int \sin(2x)dx = -\tfrac{1}{2}\cos(2x) + C$$

$$\int_0^{\frac{\pi}{2}} \sin(2x)dx = 1 \quad \text{(Fill in the details yourself)}$$

DO THIS NOW! 9.8

Do these integrals (i) - (iv) using the results in your table.

(i) $\displaystyle\int x^{1.5} dx$ 　　　　　　　　　　(ii) $\displaystyle\int e^{-2x} dx$

(iii) $\displaystyle\int_0^{\frac{\pi}{4}} 5\cos(2t)dt$ 　　　　　　(iv) $\displaystyle\int_0^{\frac{\pi}{4}} 3\sin(4t + \frac{\pi}{4})dt$

Always check your own answers using your CAS technology.

Doing integrals analytically is a set of skilful tricks or recipes, and to study every technique would go far beyond the scope of this book. We will look at just a small collection of methods: parts, partial fractions, substitution or change of variable, and use of special trigonometric function formulae.

9.7.3 Integration by parts

If the product rule for differentiation is reversed then this gives the process called integration by parts. It is mainly used for the family of integrals involving expressions like $x\cos(x)$ or $x\sin(x)$ or xe^x. The *integration by parts formula* is:

$$\int u.\frac{dv}{dx}dx = u.v - \int v.\frac{du}{dx}dx \ .$$

The idea is that by choosing u and v right you can use this formula to reduce an undo-able integral to a do-able one, possibly in your tables. Teachers (and students) tend to like this technique as it has a definite formula and set of rules, but bear in mind that it is only of use in a limited range of integrals in engineering!

It is important then to choose u and v the right way round - you can do this using the mnemonic **LATE** to choose u - choose any **L**ogarithm first, if not that then an **A**lgebraic expression, then **T**rigonometric function, then **E**xponential function. Whatever is left becomes $\dfrac{dv}{dx}$ in the formula.

The best way to see how it works is to follow this example:

Example 9.7.2

Evaluate $\int xe^x dx$.

Having first identified it as a case for integration by parts (this can be the hardest bit!), then you first need to choose u using **LATE** (LogAlgTrigExp). There are no logarithms, but there is an algebraic expression x, so:

Choose $u = x$. That means you must choose $\dfrac{dv}{dx}$ to be e^x. The original integral then corresponds to $\int u.\dfrac{dv}{dx}dx$ in the formula. To complete the use of the formula now, you need to calculate $\dfrac{du}{dx}$ and v.

You have $u = x$ and so differentiating $\dfrac{du}{dx}=1$. You have $\dfrac{dv}{dx}=e^x$ and so $v = e^x$ by simple integration.

Putting all this together into the formula,

$$\int xe^x dx = \int u.\frac{dv}{dx}dx = u.v - \int v.\frac{du}{dx}dx = xe^x - \int e^x.1dx = xe^x - e^x + C \ .$$

Note that if you choose u and v the wrong way round then you make things worse - try it and see! Also note that you can check your answer (or even avoid all this) using a CAS for indefinite or definite integrals, or numerical methods (perhaps on a spreadsheet) for definite integrals.

DO THIS NOW! 9.9

Evaluate these integrals using the parts formula and check your answers using technology (CAS or otherwise):

(i) $\int xe^{-x}dx$ (ii) $\int_0^{\pi} t\cos(t)dt$

(iii) $\int_0^{\pi} t\sin(2t)dt$ (iv) $\int \ln(x)dx$.

(Hint for (iv) - treat it as $1 \times \ln(x)$ and apply the formula.)

9.7.4 Integration by partial fractions

The technique of algebraic partial fractions allows you to split a complicated algebraic fraction into a sum of less complicated expressions, and this can help with integration. We shall restrict ourselves at this stage to dealing with relatively simple expressions of the form $\dfrac{(x-a)}{(x-b)(x-c)}$ - the method can be extended to more complicated expressions later as necessary. The technique states that this fraction can be broken up into a sum of smaller parts (partial fractions) of the form $\dfrac{(x-a)}{(x-b)(x-c)} = \dfrac{A}{(x-b)} + \dfrac{B}{(x-c)}$. The trick is to find values for A and B to make it work. First we present an example which shows the method for finding A and B. Then we use this approach to evaluate an integral.

Example 9.7.3

Split the expression $\dfrac{1}{(x-1)(x-2)}$ into partial fractions. First note that this can be split into fractions of the form $\dfrac{1}{(x-1)(x-2)} = \dfrac{A}{(x-1)} + \dfrac{B}{(x-2)}$.

To find A and B there are two approaches:

Approach 1:

Multiply both sides by the denominator of the LHS:

$$\frac{(x-1)(x-2)}{(x-1)(x-2)} = \frac{A(x-1)(x-2)}{(x-1)} + \frac{B(x-1)(x-2)}{(x-2)} .$$

Simplfy: $1 = A(x-2) + B(x-1)$.

Note that this must work for *any* value of x otherwise it is useless, so we could choose say $x = 1$ and $x = 2$ (it will be obvious why in a moment).

Choosing $x = 1$: $\qquad$ $1 = A(1-2) + B(1-1)$ and so $1 = A(-1)$ or $A = -1$.

Choosing $x = 2$: $\qquad$ $1 = A(2-2) + B(2-1)$ and so $1 = B(1)$ or $B = 1$.

Thus the fraction can be split into parts as $\dfrac{1}{(x-1)(x-2)} = \dfrac{-1}{(x-1)} + \dfrac{1}{(x-2)}$.

Approach 2: The cover up rule

This rule is very convenient but generally only works for linear factors so will not extend to more complicated cases.

We aim to find A and B in $\qquad$ $\dfrac{1}{(x-1)(x-2)} = \dfrac{A}{(x-1)} + \dfrac{B}{(x-2)}$.

The cover-up rule is this: to get what goes over $(x-1)$, choose the value of x making this 0 (i.e. $x = 1$), cover up $(x - 1)$ in the LHS and substitute $x = 1$ in the rest of the LHS. Thus $A = \dfrac{1}{(1-2)} = -1$.

Similarly for B, choose $x = 2$, and substitute this into LHS, covering up $(x - 2)$. Thus $B = \dfrac{1}{(2-1)} = 1$, agreeing with Approach 1.

Now we shall apply this to an integral.

Example 9.7.4

Evaluate $\displaystyle\int \dfrac{1}{(x-1)(x-2)} dx$.

Again the first problem is to identify that the technique you need is partial fractions. Having done that, we have just seen in Example 9.7.3 that we can write

$$\dfrac{1}{(x-1)(x-2)} = \dfrac{-1}{(x-1)} + \dfrac{1}{(x-2)} .$$

Thus the integral becomes

$$\int \dfrac{1}{(x-1)(x-2)} dx = \int \dfrac{-1}{(x-1)} + \dfrac{1}{(x-2)} dx = -\ln(x-1) + \ln(x-2) + C \text{ using tables.}$$

Note once more that you can check your answer using a CAS for indefinite integrals.

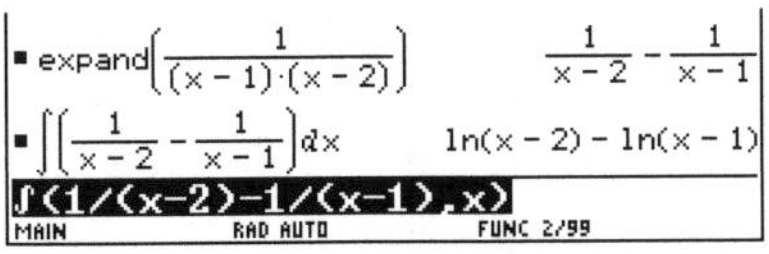

Figure 9.11 Checking your integral with CAS – using both partial fractions and integration.

There is a sophisticated issue which is perhaps worth a brief mention here, which is that there is a problem if the range of integration includes the points $x = 1$ or $x = 2$.

This is because at these points the expression to be integrated has a zero on the bottom, and division by zero is not allowed. As with any case like this, care must be taken, but full discussion is beyond the scope of this book.

DO THIS NOW! 9.10
Evaluate these integrals using partial fractions and check your answers using technology (CAS or otherwise):

(i) $\displaystyle\int\frac{1}{(x-2)(x-3)}dx$
　　　　　　　　　　　　　　　　(ii) $\displaystyle\int\frac{2}{(x-1)(x+1)}dx$

(iii) $\displaystyle\int_{0}^{1}\frac{1}{(x+1)(x+2)}dx$
　　　　　　　　　　　　　(iv) $\displaystyle\int\frac{x+1}{(x+2)(x+3)}dx$.

9.7.5 Integration by substitution or change of variable

The technique of changing variable, or substitution, is sometimes found the hardest to learn, but is unfortunately one of the most useful. It is the "undoing" of the "chain rule" for differentiation! It is useful in several different circumstances, but we shall only look at one here, which is that you can apply it to integrals of the form

$$\int(\text{something involving a function } f(x))\times\frac{df}{dx}dx \ .$$

In functional notation, we can express this rather messy form as $\int g(f(x)).f'(x)dx$. In other words you are looking for a function and its derivative to be present. This can be hard to spot unless you have good algebraic skills. You then create a new variable $u = f(x)$ and work with that.

Once more the general idea is best illustrated through an example, but really several examples are needed to show the range of this method:

Example 9.7.5

Evaluate $\int 2x\sin(x^2)dx$.

You first note that the expression $f(x) = x^2$ is there in brackets and so is its derivative $\dfrac{df}{dx} = 2x$. You therefore change the variable by letting $u = x^2$. This means that $\dfrac{du}{dx} = 2x$, and so in informal terms $du = 2xdx$. Note you have to replace everything involving x by the equivalent thing in u - it is like changing units - you must either work entirely in inches or entirely in centimetres but you cannot mix!

Thus we can replace the integral with

$$\int 2x\sin(x^2)dx = \int\sin(x^2).2xdx = \int\sin(u)du = -\cos(u)+C = -\cos(x^2)+C$$

(It is generally a good idea to change back to x at the end.)

Note once more that you can check your answer using CAS, or by taking your answer and differentiating it to see if you get back to the original expression.

DO THIS NOW! 9.11

Evaluate these integrals by change of variable and check your answers using a CAS.

(i) $\int x e^{x^2} dx$

(ii) $\int_0^\pi t \cos(t^2) dt$

(iii) $\int_0^1 \sin(x) \cos(x) dx$

(iv) $\int \frac{1}{x} \ln(x) dx$

9.7.6 Integration by use of trigonometric function formulae

Finally in this section, we note and recall from Chapter 4 that there are many relationships between trigonometric functions, and these can often allow an integral to be evaluated if used at the right moment! It is very difficult to give general rules here - we merely present a first example and then throw you in at the deep end with your list of trigonometric formulae and your fertile imagination.

Example 9.7.6

Evaluate $\int \cos^2 x dx$ (This integral is required to find root-mean-square current in an electrical circuit - see chapter 10).

Make sure you understand the notation here - $\cos^2(x)$ is a poor notation which means $\cos(x) \times \cos(x) = (\cos(x))^2$.

Recall that there are two useful formulae involving squared trigonometric functions, apart from the well known $\sin^2(x) + \cos^2(x) = 1$, which are:

$$\cos^2 A = \tfrac{1}{2}(1 + \cos(2A)) \qquad \text{and} \qquad \sin^2 A = \tfrac{1}{2}(1 - \cos(2A))$$

The first of these looks promising, and indeed the integral can be written

$$\int \cos^2 x dx = \int \tfrac{1}{2}(1 + \cos(2x)) dx = \int \tfrac{1}{2} + \tfrac{1}{2}\cos(2x) dx = \tfrac{1}{2}x + \tfrac{1}{4}\sin(2x) + C .$$

Yet again note that you can check your answer using your CAS, although it may well be the case that, as we commented earlier in the book, the corresponding answer may not look the same (e.g. Figure 9.12). In this case do not despair. You can either use your algebraic skills to show both solutions are the same, or check by differentiating the result, or perhaps even plot both results and see if their graphs are the same!

$$\int (\cos(x))^2 \, dx \qquad \frac{\sin(x) \cdot \cos(x)}{2} + \frac{x}{2}$$

Figure 9.12 Check your trigonometric integrals using a CAS.

DO THIS NOW! 9.12

Evaluate these integrals by using an appropriate trigonometric formula and check your answers using a CAS.

(i) $\displaystyle\int_0^\pi \cos^2 x\,dx$ (ii) $\displaystyle\int_0^\pi \sin^2 x\,dx$

(iii) $\displaystyle\int_0^1 2\sin(x)\cos(x)\,dx$ (iv) $\displaystyle\int_0^2 \cos(x)\cos(2x)\,dx$.

END OF CHAPTER 9 - MIXED EXERCISES – DO ALL THESE NOW!

1. Evaluate the following integrals where possible, both by hand and checking by a CAS. In each case indicate any finite values which may cause problems if the integrals become definite.

   (a) $\int x^{4/5}\,dx$ (b) $\int (1+x^2)^2\,dx$ (c) $\int \sqrt{1+x}\,dx$

   (d) $\int \dfrac{3}{2z+5}\,dz$ (e) $\int x\sqrt{x^2+1}\,dx$ (f) $\int \dfrac{1}{(x-1)(x-4)}\,dx$

   (g) $\int te^{-t}\,dt$ (h) $\int te^{t^2}\,dt$ (h) $\int e^{t^2}\,dt$

2. The figure shows a graph of a function $f(t)$ over the range $1 \le t \le 5$.

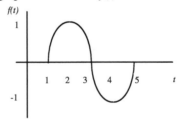

   For each of the following statements answer *always true* or *not always true*, and explain your answer in words or diagrams:

   (a) $\displaystyle\int_1^3 f(t)\,dt \approx -\int_3^5 f(t)\,dt$ (b) $\displaystyle\int_1^3 f(t)\,dt + \int_3^5 f(t)\,dt = \int_1^5 f(t)\,dt$

   (c) $\displaystyle\int_1^3 f(t)\,dt < 3$ (HINT: All true!)

3. For (a)-(d), where $f(t)$ and $g(t)$ are any functions of t, and c and a are any constants, answer *always true* or *not always true*, and explain why.

   (a) $\int c.f(t)\,dt = c\int f(t)\,dt$ (b) $\int f(t).g(t)\,dt = \int f(t)\,dt.\int g(t)\,dt$

   (c) $\int f(at)\,dt = \int a.f(t)\,dt$ (d) $\int \dfrac{f(t)}{g(t)}\,dt = \dfrac{\int f(t)\,dt}{\int g(t)\,dt}$

   (HINT: Only (a) is true.)

10

Integral Calculus 2

"Mathematics consists in proving the most obvious thing in the least obvious way." George Polya

10.1 INTRODUCTION

In Chapter 9, you met the idea of integration in its two aspects: as a summation process and as the inverse process of differentiation. In this chapter, you will have a chance to explore both those ideas a little further. You will begin by going further into the summation process as it applies to finding mean and root-mean-square (RMS) values, electrical charging processes, volumes of revolution and surface areas.

Finally, you will meet some very simple ordinary differential equations, an area of crucial importance in engineering systems, and will see how integration can be used to solve them in these simple cases. These important ideas are picked up and developed in Chapters 15 and 16.

Once more, remember to **do the exercises and activities**.

10.2 INTEGRATION AS SUMMATION: MEAN AND RMS
10.2.1 The mean of a function

The mean of function over a given range in some sense describes the average, or middle value, of the function - or the value around which it is balanced. This idea is illustrated in Figure 10.1.

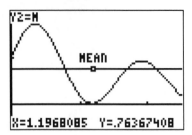

Figure 10.1 Graphic representation of the mean of a function over a range.

The standard definition of the mean of a set of y values is $\dfrac{\text{sum of } y \text{ values}}{\text{number of values}}$.

This definition would be fine if y were only defined at a finite number of points, as in Figure 10.2. In that case, we can write:

$$\text{mean} = \bar{y} = \frac{\sum\limits_{i=1}^{n} y_i}{n} \quad \text{which can be expressed as} \quad \bar{y} = \frac{\sum\limits_{i=1}^{n} y_i}{\sum\limits_{i=1}^{n} 1} = \frac{\sum\limits_{i=1}^{n} y_i}{\sum\limits_{i=1}^{n} 1} \times \frac{\delta x}{\delta x} = \frac{\sum\limits_{i=1}^{n} y_i \, \delta x}{\sum\limits_{i=1}^{n} 1 \delta x}.$$

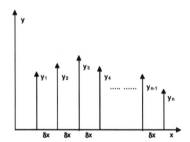

Figure 10.2 Representing a set of discrete values.

However, this is not the case if y is defined as a continuous range of values - an infinite number of values across a finite range. How are we to deal with this? We give only a very brief and non-rigorous description here. Just as with areas in Chapter 9, the number of y strips increases and the gap δx between them decreases. Once more, each of the summations becomes an integral between limits a and b, with δx becoming the dx at the end of the integral.

Thus, the mean of a function $y(x)$ over a range $a \le x \le b$ is given by the formula

$$\bar{y} = \underset{dx \to 0}{Lim} \frac{\sum\limits_{i=1}^{n} y_i \, \delta x}{\sum\limits_{i=1}^{n} 1 \delta x} = \frac{\int\limits_{a}^{b} y(x)dx}{\int\limits_{a}^{b} 1 dx} = \frac{\int\limits_{a}^{b} y(x)dx}{b-a}.$$

That is, the mean over the interval is $\quad \bar{y} = \dfrac{\displaystyle\int_a^b y(x)dx}{b-a}$.

Intuitively the formula $\bar{y} = \dfrac{\text{sum of } y \text{ values}}{\text{number of values}}$ becomes $\bar{y} = \dfrac{\text{integral of } y \text{ values}}{\text{length of interval}}$.

Example 10.2.1

Plot the function $y = t$ over the range $0 \le t \le 1$. Evaluate the mean of this function over that same range and mark it on your diagram to see if it looks right.

The mean over $0 \le t \le 1$ is $\quad \bar{y} = \dfrac{\displaystyle\int_0^1 t\,dt}{1-0} = [\tfrac{1}{2}t^2]_0^1 = \tfrac{1}{2}$.

Figure 10.3 shows the plot of the function, with mean value marked, and it does indeed look sensible.

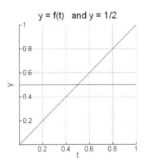

Figure 10.3 "Seeing" whether your mean value is right.

DO THIS NOW! 10.1

Use the formula to work out the mean of the waveform $y = \sin(t)$ over the range $0 \le t \le \pi$. Draw a picture to see if your answer is sensible. Next, by considering the symmetry of the situation (draw a picture again), *predict* the mean of this function over the range $\pi \le t \le 2\pi$. Check your prediction by working out the mean using the formula. Finally, *predict* the mean of the sine wave over the range $0 \le t \le 2\pi$, and verify your answer using the formula.

Repeat all the parts of this exercise for the so-called half-wave rectified sine wave (important in electronics) given by

$$y = \begin{cases} \sin(t) & 0 < t \le \pi \\ 0 & \pi < t \le 2\pi \end{cases}$$

$y(t+2\pi) = y(t)$ (i.e. repeating every 2π seconds)

Once more, predict answers where possible, before calculating them.

10.2.2 The Root-Mean-Square of a function

The mean does not give a good measure of the average *size* of a function, in the sense of "how big" the function is, regardless of sign. The measure used to do this is the root-mean-square (RMS), defined by:

$$y_{RMS} = \sqrt{\frac{1}{b-a} \int_a^b (y(t))^2 \, dt}.$$

The RMS is (the square **R**oot of (the **M**ean of (the **S**quared values of *y*))).

This rather complicated looking definition is used because squaring *y* removes the effect of any signs and just measures the size. Also it can have physical meaning. In electrical theory, for example, it is related to power consumption.

Example 10.2.2

Evaluate the RMS of the function $y = t$ over the range $0 \le t \le 1$. Plot the function over that range and mark the RMS on your diagram to see if it looks right. Also, mark the mean as calculated earlier. Are they the same? Are they close? What would you expect here?

The RMS over $0 \le t \le 1$ is $\sqrt{\frac{1}{1-0} \int_0^1 t^2 \, dt} = \sqrt{[\frac{1}{3} t^3]_0^1} = \sqrt{\frac{1}{3}} \approx 0.577$.

Figure 10.4 shows the plot of the function, with both RMS and mean value marked. It does indeed look sensible, and you should note that mean and RMS do not give the same answers, even over ranges where the function is always positive.

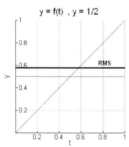

Figure 10.4 Graph showing both mean and RMS of a function.

DO THIS NOW! 10.2

Use the RMS formula to work out the RMS of the waveform $y = \sin(t)$ over the range $0 \le t < \pi$. Draw a picture to see if your answer is sensible. Compare with the mean over the same range. Next (draw a picture again), predict the RMS of this function over the range $\pi \le t \le 2\pi$. Check your prediction by working out the RMS using the formula. Next predict the RMS of the sine wave over the range $0 \le t \le 2\pi$, and verify your answer using the formula.

Repeat all the parts of this exercise for the half-wave rectified sine wave as defined in **DO THIS NOW! 10.1** Comment on differences between your results here and there.

10.3 INTEGRATION AS SUMMATION: CHARGE ACCUMULATION

You have already met the relationship $i = \dfrac{dq}{dt}$ between charge $q(t)$ and current $i(t)$. Turning the relationship the other way round, you get $q = \int i\, dt$. Here is an example of the application of this formula.

Example 10.3.1

An initially uncharged capacitor is charged through a resistor so that the current flowing at time t is $i(t) = 0.03e^{-3t}$. Find the charge $q(t)$ on the capacitor at time t.

The indefinite integration gives (with an arbitrary constant of integration C):

$$q(t) = \int i\, dt = \int 0.03e^{-3t}\, dt = -0.01e^{-3t} + C \ .$$

To find C you can use the fact that the capacitor initially holds no charge, that is when $t = 0$, $q = 0$. Putting this into the solution

$$0 = -0.01 + C, \text{ and so } C = 0.01.$$

Thus, the full solution is $q(t) = 0.01(1 - e^{-3t})$.

We could ask supplementary questions.

What is the charge after 0.2 seconds? Simply put t equal to 0.2 in the formula.

What is the eventual value of the accumulated charge after a "long time"? (a "long time" here just means "long enough for things to settle down".) Note that after a long time e^{-3t} approaches 0, so the eventual settled level of charge is 0.01.

The above approach used an *indefinite* integral. There is an alternative approach using a *definite* integral, which gives the same result. The charge accumulating between times t_1 and t_2 is given by the definite integral

$$charge = q = \int_{t_1}^{t_2} i\, dt \ .$$

Example 10.3.2

In a particular circuit, the current is given by $i(t) = 2\sin(t)$. Using the definite integral formula, find the charge accumulating between the times $t = 0$ and $t = \pi$.

$$\text{Charge} = q = \int_0^{\pi} 2\sin(t)\, dt = \left[-2\cos(t)\right]_0^{\pi} = -2\cos(\pi) - (-2\cos(0)) = 4 \ .$$

DO THIS NOW! 10.3

In Example 10.3.2, what would you predict to be the charge accumulating between the times $t = 0$ and $t = 2\pi$? Explain your prediction and check if you are right by doing the integral.

10.4 INTEGRATION AS SUMMATION: VOLUME AND SURFACE AREA

The set of ideas used to justify the fact that the area under a curve (see 9.4) can be found by an integral can be used much more widely, as indeed in 10.2 above. We saw there, and earlier, the application of the idea that you can deal with finding the area under a continuous function by treating the area as a set of vertical strips or elements, and then realising that when the number of strips gets larger with each individual strip getting narrower, the end of that process is that the sum of the strip areas (that is, the total area under the curve) approaches the value calculated by the definite integral.

This idea can be extended to find volumes of revolution, as shown in Figure 10.5. In this figure, a curve defined by $y = f(x)$ in the range $a \le x \le b$ is rotated about the x-axis to create a so-called volume of revolution, as shown. In engineering terms, such a body could, for example, be part of a gas storage vessel, and the question can arise of how to calculate the volume enclosed by this cylindrically-symmetric body.

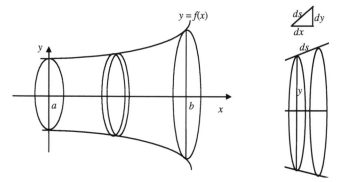

Figure 10.5 A volume of revolution.

The clue lies in the diagram. Instead of splitting an area into thin vertical strips, we split the volume into thin radial slices. The radius of one of these slices, or disks, is $f(x)$ on one side, and the thickness is δx, so its volume is approximately (using the formula for volume of a cylinder) $\delta V \approx \pi (f(x))^2 \delta x$. This is only approximately equal because actually the disk will have a slightly differing radius between one side and the other, but as we take more and more disks and each disk is thinner and thinner, this difference will matter less and less.

The total volume is found by adding up all the volumes of the disks $V \approx \sum_{all\ disks} \pi y^2 \delta x$, and if we now go through that process of slicing more and more finely so the disks become thinner and thinner, the summation becomes an integral, δx becomes dx, and the "approximately equals" becomes an "exactly equals":

$$V = \int_a^b \pi y^2 dx$$

Example 10.4.1

The curve $y = \frac{x}{4}$ between $x = 0$ and $x = 1$ is rotated around the x-axis. What size volume is generated?

Using the formula: $V = \int_0^1 \pi y^2 \, dx = \int_0^1 \pi(\frac{x}{4})^2 \, dx = \frac{\pi}{16} \approx 0.20$.

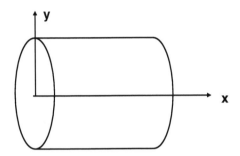

Figure 10.6 The volume of revolution in Example 10.4.1.

The picture in Figure 10.6 shows that in this example we are actually making a cylinder, of height 1 and radius $\frac{1}{4}$, and the formula for volume of a cylinder says the volume should be

$$V = \pi r^2 h = \pi(\frac{1}{4})^2 . 1 = \frac{\pi}{16} .$$

This agrees with what we have found from the integral approach.

DO THIS NOW! 10.4

Apply geometrical arguments as above to show that the curved surface area of a volume of revolution carved out by rotating $y(x)$ for values of x between $x = a$ and $x = b$ around the x-axis is given by the formula:

$$A = \int_a^b 2\pi y \sqrt{1 + (\frac{dy}{dx})^2} \, dx .$$

Use this formula to deduce that the curved surface area of the body generated in Example 10.4.1 is $\frac{\pi}{2}$.

Does this accord with what you would calculate using the standard formula for the curved surface area of a cylinder?

(HINT: $Area = 2\pi . rh$.)

DO THIS NOW! 10.5

To do this exercise you will need to get a polystyrene cup of the sort commonly used to serve tea or coffee. This is not a purely academic exercise as the same kind of shape occurs for example in modelling buckets of molten steel. The object is to calculate the volume of the cup. Start by aligning the cup with the co-ordinate axes as shown in Figure 10.7:

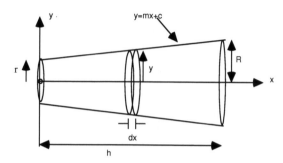

Figure 10.7 A coffee cup as a volume of revolution.

You see that you can regard the cup as a body formed by rotating the line $y = mx + c$ around the x-axis as shown. The volume of revolution will be given by the integral

$$V = \int_a^b \pi y^2 dx = \int_a^b \pi (mx+c)^2 dx \ .$$

Now, using a ruler, measure the radii of the top (R) and the bottom (r), and measure the height h. Hence find the values of m and c for your cup. If you cannot do this read this hint - but have a go first!

(Hint: the value of y intercept "c" is r; the value of slope "m" is $\dfrac{R-r}{h}$)

You also need the limits "a" and "b" for the integral. What are these?

(Hint: these are the x values at the bottom and top of the cup: $a = 0$ and $b = h$.)

Write down the integral you need to evaluate to find the volume and evaluate this integral by more than one means - CAS, graphic calculator, by hand, etc.

Is your answer reasonable? How can you check? Think NOW before reading on ...

(Hint: Here are some suggestions for checking:

- The volume must lie between that of the outer cylinder of height h and radius R, and that of the inner cylinder of height h and radius r.   (Why?)

- Perhaps look at the bottom of the cup and read what is there!)

You can now extend this in several ways, for instance by finding the curved surface area and checking how sensitive the answer is to the accuracy of your measurements.

10.5 A FIRST LOOK AT DIFFERENTIAL EQUATIONS

Now we move away from integration as summation and have a first look at how the methods of integration can be used to solve so-called differential equations. In simple terms these are equations which involve not only a function you don't know but its derivatives too. These equations are very important because many laws governing physical systems arise from describing not variable quantities themselves, but their rates of change - such as acceleration. This is only a brief introduction at this stage - you will see much more of these in Chapters 15 and 16.

In some simple cases you can solve these equations just by realising that to "undo" a differentiation you just do an integral!

Example 10.5.1

A simply supported beam rests on two supports 2 metres apart, one at either end of the beam, with both supports at a height 0. The vertical displacement of the beam, at distance x from one end, is given by y, the beam profile, which has slope given by the equation

$$\frac{dy}{dx} = 0.002x - 0.002 .$$

Find the beam profile $y(x)$.

This is an example of a simple ordinary differential equation - it is differential because it involves the derivative of the function you want; it is ordinary because it only involves ordinary (and not partial) derivatives. Solving it involves finding $y(x)$ in terms of x. In this simple case, you can solve the equation by reverse-differentiating (i.e. integrating) both sides, using the fact that integration is the inverse of differentiation:

$$y = \int (0.002x - 0.002)dx = 0.001x^2 - 0.002x + C .$$

Note that an arbitrary constant C has appeared because this is an indefinite integral. To find that constant you need an extra bit of information - in this case that at the left-hand end when $x = 0$, the beam is at height 0 and so $y = 0$. Putting this in:

$$0 = 0.001 \times 0^2 - 0.002 \times 0 + C \qquad \text{or} \qquad C = 0.$$

Thus the profile is $y = 0.001x^2 - 0.002x .$

A plot of this (Figure 10.8) shows that the answer we have predicts a sensible beam shape.

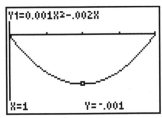

Figure 10.8 The beam shape in Example 10.5.1.

DO THIS NOW! 10.6

Given the differential equation $\dfrac{d\theta}{dt} = 2e^{-0.1t}$ and a "kick off" set of conditions that

$\theta = 10$ when $t = 0$, find the time t taken for θ to reach within 1% of its eventual steady state. We make it about 42 seconds. What do you get?

END OF CHAPTER 10 EXERCISES – DO ALL OF THESE NOW!

1. (a) Find the area enclosed between the curve $y = x^2$ and the lines $y = 0$ and $x = 1$.

   Find the mean and RMS values of $y = x^2$ over this range.

   (b) Find the area enclosed between the curves $y = x^2$ and $y = x^{\frac{1}{2}}$. (Draw a picture!)

2. A hollow casing is formed by rotating the curves $y = 2 - x^2$ and $y = 5 - x^2$ between $x = 0$ and $x = 1$ about the axis of x. Find its volume. (Hint: Find the volume generated by the outer curve first, then by the inner curve, then subtract them.)

3. In finding the bending moments at points on a beam, it was found that $M = 100x - 5x^2 + A$ where M = bending moment, x = distance from one end of the beam and A is a constant. Also $M = 520$ when $x = 10$.

   Find A. Find also the area included between the curve $M = 100x - 5x^2 + A$, the axis of x, and the lines at $x = 0$ and $x = 20$.

11

Linear Simultaneous Equations

"Mathematical concepts may be communicated easily in a format which combines visual, verbal, and symbolic representations in tight coordination." Ralph Abraham

11.1 THE NEED FOR LINEAR SIMULTANEOUS EQUATIONS

Many engineering systems need to be described by more than one variable. It might be something as simple as specifying the position of an object on a table - which needs two co-ordinates. It might be a simple electronic circuit such as the one shown in Figure 11.1 below, where the state of the circuit is described by the two electric currents i_1 and i_2.

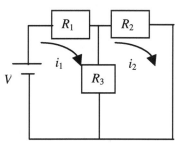

Figure 11.1 A simple network of resistors.

Often, understanding the system involves solving equations to find the values of the variables. In general, to find 2 variables with unique values you need 2 independent equations; for 3 variables you need 3 equations and so on. It is of course a little more complicated than this, and it is this mathematical idea of solving so called simultaneous equations that you will study in this chapter.

The ideas involved in solving linear simultaneous equations lend themselves nicely to both graphical and numerical/algebraic approaches, particularly when only two variables are involved. Even when more than 2 variables are involved the ideas developed with just 2 variables provide a useful framework for thinking about these problems. In this chapter we shall restrict ourselves to just 2 variable cases.

Before going on we shall answer two questions. Why do we call them *linear* simultaneous equations? The answer is that we shall only be looking at equations of the form $ax + by = c$ for constants a, b and c. Such an equation can be rewritten in

the form $y = -\dfrac{a}{b}x + \dfrac{c}{b}$, which you can see is the equation of a straight line - hence

linear. Linear equations are well understood. If we extend to *non-linear* equations then life becomes significantly more complicated!

The second question is this. What do we mean by solving a set of linear simultaneous equations? The answer is simply (for example in the case of two equations involving unknowns x and y) that we need to find values of x and y which simultaneously make both equations true.

As you work through this chapter, keep on **doing the exercises and activities**.

11.2 WHERE SIMULTANEOUS EQUATIONS OCCUR – TWO EXAMPLES
Example 11.2.1 A simple circuit

Figure 11.1 shows a simple resistor network circuit, with unknown currents i_1 and i_2.

Application of Kirchoff's first and second laws shows that the currents i_1 and i_2 must satisfy the linear simultaneous equations

$$(R_1 + R_3)i_1 - R_3i_2 = V$$
$$- R_3i_1 + (R_2 + R_3)i_2 = 0$$

Example 11.2.2 Modelling a road bend

Figure 11.2 shows a segment of road from above.

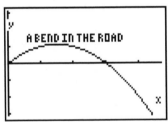

Figure 11.2 A section of a bend in a road viewed from above.

The road engineers wish to model this curve with a quadratic equation of the form $y = ax^2 + bx$. They specify that it must pass through the points (0, 0) (already satisfied); (1, 1) and (2, 1). Substituting these (x, y) values in turn into the quadratic equation gives:

$$1 = a1^2 + b1$$
$$1 = a2^2 + b2$$

Thus in order for the quadratic expression to fit the curve, the values a and b must be the solutions of the simultaneous equations

$$a + b = 1$$

$$4a + 2b = 1$$

11.3 SOLVING SIMULTANEOUS EQUATIONS GRAPHICALLY

When there are only two simultaneous equations in two unknowns, there are powerful ways of representing the concepts and solutions graphically. Perhaps the easiest way to introduce these graphical notions underlying linear simultaneous equations is to present an example.

Example 11.3.1

Consider the set of two equations for the two unknown variables x and y.

$$4x + 2y = 8 \qquad\qquad (11.1)$$
$$8x - 5y = -2 \qquad\qquad (11.2)$$

With a few simple manipulations (check our working!), these two equations (11.1) and (11.2) can be re-written in the form:

$$y = -2x + 4 \qquad\qquad (11.3)$$
$$y = \frac{8}{5}x + \frac{2}{5} \qquad\qquad (11.4)$$

These equations (11.3) and (11.4) carry exactly the same information as (11.1) and (11.2). It is just that now you may recognise in this form that each of the equations represents a straight line of the form $y = mx + c$.

Thus if a pair of values (x, y) makes one equation true, then the point (x, y) lies on the corresponding line when we draw it. If it makes the other equation true, then the point (x, y) lies on that other line when we draw it. Finally the key point: if a pair of values (x, y) makes both equations true simultaneously, then the point (x, y) lies on both lines when we draw them. That is, (x, y) is the point where the lines cross each other - the point of intersection of the lines. This also means the pair of values (x, y) form the solution of the original simultaneous equations.

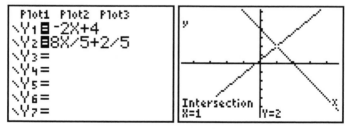

Figure 11.3 Solving Example 11.3.1 graphically.

Continuing with the example, Figure 11.3 shows the two straight lines (11.3) and (11.4) on the same diagram. It is apparent from the diagram that they cross at the point given approximately by $x = 1$ and $y = 2$.

DO THIS NOW! 11.1

Plot these two lines on your own favourite piece of graphical technology and set the window so that you find the intersection point. Zoom in on the graph until you have convinced yourself that to at least 3 decimal places the intersection point is actually (1.000, 2.000).

Now substitute these values back into the original equations (11.1) and (11.2) and convince yourself that the values $x = 1$ and $y = 2$ really do satisfy those original equations - that is, they are the solution to those equations.

11.4 SOLVING SIMULTANEOUS EQUATIONS WITH SIMPLE NUMBERS

The graphical method can be backed up by simple tabulation of the y values coming from both equation (11.3) and equation (11.4), followed by inspection of the table for places where the (x, y) values coincide, which again gives the point of intersection

This is illustrated for the equations of Example 11.3, in Table 11.1 below.

Table 11.1 Solving Example 11.3.1 in a table.

11.5 SOLVING SIMULTANEOUS EQUATIONS ALGEBRAICALLY

If you have met simultaneous equations before (which you almost certainly have in some form or other) then you probably will have used algebraic methods to solve them. There are various approaches to this, but we shall look at just two of them.

11.5.1 Elimination of variables

The numbers in the equations will affect how easy this approach is to apply. We shall illustrate a typical solution strategy with equations (11.1) and (11.2) again.

$$4x + 2y = 8 \qquad\qquad (11.1)$$
$$8x - 5y = -2 \qquad\qquad (11.2)$$

A possible strategy is to eliminate x from the equations by taking twice equation (11.1) away from equation (11.2). Thus

(11.2) - 2×(11.1) $(8 - 2\times4)x + (-5 - 2\times2)y = (-2 - 2\times8)$

which simplifies to $-9y = -18$ or $y = 2$ as before.

To find x either eliminate y similarly, or substitute the value of y back into either equation and solve that to find x. E.g. substituting back into equation (11.1) gives:

$$4x + 2*(2) = 8 \quad \text{which is solved to give } x = 1.$$

11.5.2 Cramer's rule and determinants

A more systematic algebraic way of solving simultaneous equations is Cramer's rule. This is a little old-fashioned as an algorithm for actually solving equations, but is useful because it is a way of introducing and emphasising some important general points which will help you to understand simultaneous equations; such as the connection between the shape of the equations and the way they are written, and the crucial role played by the important notion of the ***determinant***.

It will be helpful to introduce a general way of representing equations, which relates the variable and constant names to their position in the equations, and which will expand easily as we go to more equations, without running out of letters to use. Thus the general set of two equations in the two unknowns x_1 and x_2 can be written

$$a_{11}x_1 + a_{12}x_2 = b_1 \qquad\qquad (11.5)$$

$$a_{21}x_1 + a_{22}x_2 = b_2 \qquad\qquad (11.6)$$

Note that the constant coefficient names are subscripted - that is x_1 is multiplied by a_{11} and so on. The coefficient subscripts are chosen to represent the position in the equations, in the form of row number followed by column number. That is, a_{11} is in row 1 column 1, a_{12} is in row 1 column 2, and so on. It is helpful to write this in *matrix form* as in equation (11.7) (see chapter 12 for what a matrix is in general).

$$\begin{bmatrix} a_{11} & a_{12} \\ a_{21} & a_{22} \end{bmatrix} \begin{bmatrix} x_1 \\ x_2 \end{bmatrix} = \begin{bmatrix} b_1 \\ b_2 \end{bmatrix} \qquad\qquad (11.7) \ .$$

This form helps you to see the essential shape of the equations - as long as they were written neatly and systematically in the standard form of (11.5) and (11.6) in the first place! To see how this helps, we shall go on and generate Cramer's rule.

You can, with some algebraic pain, or maybe using technology as in figure 11.4, solve equations (11.5) and (11.6) to give the general solution of any pair of linear simultaneous equations. We shall spare you the details and just present the solutions.

$$x_1 = \frac{b_1 a_{22} - b_2 a_{12}}{a_{11}a_{22} - a_{21}a_{12}}$$
$$\qquad\qquad (11.8)$$
$$x_2 = \frac{a_{11}b_2 - a_{21}b_1}{a_{11}a_{22} - a_{21}a_{12}}$$

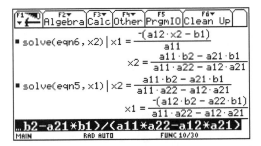

Figure 11.4 Solving simultaneous equations using CAS (TI-92).

This is messy, and difficult to read and remember, but it can be made clearer using the idea of a determinant. You might notice that without getting too detailed you can see that all sections of the solutions have the general form $(a.d - c.b)$. This prompts us to define a thing called a two by two (2×2) determinant, as

$$\begin{vmatrix} a & b \\ c & d \end{vmatrix} = (a.d - c.b)$$

Note two things here.

1. The determinant is written with *straight lines* down either side, not brackets.

2. Although it is laid out as an array, it is *just a number* - calculated by

(top left × bottom right -bottom left × top right).

Now to see why we have defined this in this way, let's return to the general solution formulae from above.

The equations in matrix form (11.7) were

$$\begin{bmatrix} a_{11} & a_{12} \\ a_{21} & a_{22} \end{bmatrix} \begin{bmatrix} x_1 \\ x_2 \end{bmatrix} = \begin{bmatrix} b_1 \\ b_2 \end{bmatrix}$$

with solutions from (11.8)

$$x_1 = \frac{b_1 a_{22} - b_2 a_{12}}{a_{11} a_{22} - a_{21} a_{12}}$$

$$x_2 = \frac{a_{11} b_2 - a_{21} b_1}{a_{11} a_{22} - a_{21} a_{12}}$$

These solutions can now be written in terms of determinants, and when you do that, you get is what is called **Cramer's Rule**:

$$x_1 = \frac{\begin{vmatrix} b_1 & a_{12} \\ b_2 & a_{22} \end{vmatrix}}{\begin{vmatrix} a_{11} & a_{12} \\ a_{21} & a_{22} \end{vmatrix}} \qquad x_2 = \frac{\begin{vmatrix} a_{11} & b_1 \\ a_{21} & b_2 \end{vmatrix}}{\begin{vmatrix} a_{11} & a_{12} \\ a_{21} & a_{22} \end{vmatrix}} \qquad (11.9)$$

Note that Cramer's rule (11.9) is easier to remember than the general form (11.8), because its expression in determinant form is related to the shape of the equations:

1. The determinant on the *bottom* of each solution is that coming from the left-hand-side (LHS) matrix of the equations.

2. The determinant on the top of the *first* variable solution is the LHS, but with column *one* replaced by the right-hand-side (RHS) matrix in the equations.

3. The determinant on the top of the *second* variable solution is the LHS, but with column *two* replaced by RHS in the equations.

Example 11.3.1 revisited (again!)

Solve the equations of Example 11.3.1 again, but this time using Cramer's rule.

$$4x + 2y = 8 \qquad (11.1)$$

The equations are

$$8x - 5y = -2 \qquad (11.2)$$

Renaming the variables to x_1 and x_2, these can be written in matrix form as

$$\begin{bmatrix} 4 & 2 \\ 8 & -5 \end{bmatrix} \begin{bmatrix} x_1 \\ x_2 \end{bmatrix} = \begin{bmatrix} 8 \\ -2 \end{bmatrix} \qquad (11.10)$$

Cramer's rule applied to (11.10) then says:

$$x_1 = \frac{\begin{vmatrix} 8 & 2 \\ -2 & -5 \end{vmatrix}}{\begin{vmatrix} 4 & 2 \\ 8 & -5 \end{vmatrix}} \qquad\qquad x_2 = \frac{\begin{vmatrix} 4 & 8 \\ 8 & -2 \end{vmatrix}}{\begin{vmatrix} 4 & 2 \\ 8 & -5 \end{vmatrix}} \qquad (11.11)$$

Simplifying the determinants in (11.11) then gives the same answers as before:

$$x_1 = \frac{-40 - (-4)}{-20 - (16)} = \frac{-36}{-36} = 1 \qquad\qquad x_2 = \frac{-8 - (64)}{-20 - (16)} = \frac{-72}{-36} = 2.$$

DO THIS NOW! 11.2

Solve the following sets of equations using Cramer's rule and check your answer by algebraic elimination, by geometrical solution, by tabulation, and by substituting back into the original equations to see if the answers work!

(a) $\begin{aligned} x + 3y &= 7 \\ x + y &= 3 \end{aligned}$ 　　　　(b) $\begin{aligned} a + 5b &= 21 \\ a + 2b &= 9 \end{aligned}$

(c) $\begin{aligned} x_1 - 2x_2 &= -4 \\ 3x_1 + x_2 &= 15 \end{aligned}$ 　　　　(d) $\begin{aligned} -\alpha + 5\beta &= 0 \\ \alpha + 2\beta &= 9 \end{aligned}$.

11.6 SOLVING SIMULTANEOUS EQUATIONS USING TECHNOLOGY

Your various technologies will have a variety of ways for solving simultaneous equations. Some will have a direct solver; some will require you to use matrix theory (see Chapter 12). Some will be able to handle general algebraic equations but others will be restricted to just numbers.

DO THIS NOW! 11.3

Find out NOW how your favourite piece of technology, graphic calculator or a PC package, deals with linear simultaneous equations. Does it have a direct solver built in? If not, what is available? (If you must use a matrix inverse then you need to go on to Chapter 12 here before being able to use your solver!). Does it have computer algebra capability or will it only deal with numbers? Will it deal with complex numbers as coefficients?

Check out whether you can use your technological solver, by verifying your answers to the exercises in DO THIS NOW 11.2!

> **DO THIS NOW! 11.4**
>
> Check that you have understood so far, by solving the following sets of equations:
>
> 1. Solve them all by technology.
>
> 2. Solve them all algebraically, by elimination and/or Cramer's rule.
>
> 3. Solve them all graphically.
>
> 4. Check your answers by substituting them back into the *original* equations and making sure they make the LHS = RHS.
>
> (a)
> $$x + y = 12$$
> $$x - y = 6$$
>
> (b)
> $$2x + y = 10$$
> $$x - y = 2$$
>
> (c)
> $$4x + y = 10$$
> $$3x + y = 9$$
>
> (d)
> $$2x + 3y = 11$$
> $$4x + y = 12$$

11.7 EQUATIONS WITH NO UNIQUE SOLUTION - SINGULAR EQUATIONS

Things are not always as simple as you have seen so far. Consider this example:

Example 11.7.1 Singular equations with no solutions

Try to solve the simultaneous equations
$$2x_1 + x_2 = 3 \qquad (11.12)$$
$$4x_1 + 2x_2 = 10 \qquad (11.13)$$

You should find you hit a problem. For instance if you try to eliminate x_1 by doing

(11.13) - 2*(11.12), you get: $\qquad 0 = 4$.

Both variables have disappeared and we have an impossible equation!

(11.12) and (11.13) form an example of what we call a *singular* set of equations. The problem may be thought of in different ways:

- The two equations are incompatible. To see this, divide (11.13) by 2:

$$2x_1 + x_2 = 3 \qquad (11.12)$$
$$2x_1 + x_2 = 5 \qquad (11.14)$$

These cannot both be true simultaneously since the left hand sides are identical now and must either equal 3 or 5, but cannot equal both.

- Algebraically, if we try to use Cramer's rule, then the solution for both variables involves dividing by the determinant of the LHS. In this case that determinant is

$$\begin{vmatrix} 2 & 1 \\ 4 & 2 \end{vmatrix} = (2\times2 - 4\times1) = 0 \text{ , and since this is 0 you cannot divide by it.}$$

- Geometrically, you can rewrite the two equations (11.12) and (11.13) in straight line form (note that x_2 is "y" and x_1 is "x" when we plot) and try to find where they intersect:

(11.12)
$$x_2 = -2x_1 + 3$$

(11.13)
$$x_2 = \frac{-4x_1 + 10}{2}$$

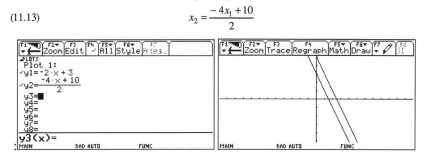

Figure 11.5 Getting a graphical understanding of singular equations.

The plots appear in Figure 11.5. It is apparent that the lines look (and indeed are) parallel, so they never meet, they have no point of intersection, and the equations therefore have no simultaneous solution.

There is a variation on this problem, illustrated in Example 11.7.2.

Example 11.7.2 Singular equations with an infinite number of solutions

Try now to solve the simultaneous equations
$$2x_1 + x_2 = 5 \qquad (11.15)$$
$$4x_1 + 2x_2 = 10 \qquad (11.16)$$

Again you can see that the determinant of the LHS is 0, so you know there will be a problem - which algebraically will show itself by a zero determinant in Cramer's rule. What is interesting here is to see what is happening geometrically.

Rewrite the two equations (11.15) and (11.16) in standard straight line form:

(11.15)
$$x_2 = -2x_1 + 5$$

(11.16)
$$x_2 = \frac{-4x_1 + 10}{2} = -2x_1 + 5$$

You can see that actually both equations represent the same straight line and will therefore meet at *every* point along their length - there are *infinitely* many solutions - any point on the line! In fact there is really only one independent equation here at all, so the rule that you need two independent equations to give a unique solution to two unknowns is broken.

Singular and non-singular equations

In general if the determinant of the LHS is *not* zero, then a set of equations will have a unique solution, and we call the system *non-singular*. If the determinant of the LHS *is* zero, then a set of equations will have either no solutions or an infinite number of solutions, and we call the system *singular*.

DO THIS NOW! 11.5

A set of simultaneous equations with unknowns x and y and constant α, is defined in matrix form as follows:

$$\begin{pmatrix} 3 & 1 \\ 6 & 2 \end{pmatrix}\begin{pmatrix} x \\ y \end{pmatrix} = \begin{pmatrix} 1 \\ \alpha \end{pmatrix}.$$

Explain in your own way, both algebraically and graphically, why it is not possible to find a unique solution to the pair of equations, for each of the two cases (a) $\alpha = 2$, and (b) $\alpha = 4$.

11.8 ILL CONDITIONED EQUATIONS

Sometimes numbers do not behave well, even with linear equations. We are used to rounding answers to a certain number of decimal places, but in certain circumstances such rounding can be part of a big problem, called ill-conditioning.

With certain sets of simultaneous equations, in certain circumstances, the solutions can be very sensitive indeed to any rounding of the coefficients, as illustrated in Example 11.8.1. You will see that the problem arises when the determinant of the LHS is not exactly zero, but is somewhere "near" to zero.

Example 11.8.1 An example of ill-conditioned simultaneous equations

Let us try to solve equations (11.17) and (11.18) for various values of constant α.

$$2x_1 + x_2 = 3 \qquad (11.17)$$
$$\alpha x_1 + 2x_2 = 10 \qquad (11.18)$$

First note that if $\alpha = 4$ *exactly* then the equations are singular and there are no solutions.

Now we solve them for (a) $\alpha = 4.01$, and (b) $\alpha = 3.99$. That is, we make just a very small change in α and see what effect it has.

Case $\alpha = 4.01$

Equations are
$$2x_1 + x_2 = 3 \qquad (11.17)$$
$$4.01x_1 + 2x_2 = 10 \qquad (11.19)$$

and the solutions (check it out with your favourite method or technology) are
$$x_1 = 400 \qquad\qquad x_2 = -797 .$$

Case $\alpha = 3.99$

Equations are
$$2x_1 + x_2 = 3 \qquad (11.17)$$
$$3.99x_1 + 2x_2 = 10 \qquad (11.20)$$

and the solutions (again check it out with your favourite method or technology) are
$$x_1 = -400 \qquad\qquad x_2 = 803 .$$

Now it is time to reflect. The first point as you compare the solutions of the two cases is that you may wonder whether you have made a mistake with the signs - a reasonable thing to check - check it out and you will find that you have not!

Next, look at what has happened. One coefficient in the equations (α) has changed by a very small percentage (approximately 0.5%) - which is a very reasonable error in engineering terms. But the solutions for x_1 and x_2 have changed enormously, by about 200%.

How can we understand what is happening here? Have a look at the problem geometrically. Equations (11.16) and (11.17) are translated into straight line form in equations (11.21) and (11.22). Figure 11.6(a) shows the corresponding geometrical representation of these equations, for the case $\alpha = 4.01$; Figure 11.6(b) shows that for the case $\alpha = 3.99$.

$$x_2 = -2x_1 + 3 \qquad (11.21)$$

$$x_2 = \frac{-\alpha}{2}x_1 + \frac{10}{2} \qquad (11.22)$$

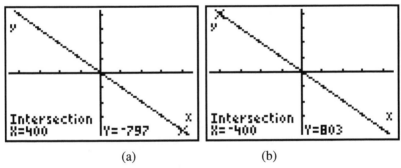

(a)   (b)

Figure 11.6 The drastic effect of ill-conditioning.

It is apparent that in each case the two lines are very nearly parallel to each other. This shouldn't be surprising. If the LHS determinant had been exactly zero they would be exactly parallel, so as it is nearly zero they are nearly parallel. Thus the point where they meet and cross will be very sensitive to the slope of each line. A small change in slope can move the intersection point from one place to another one far away. In Figure 11.6(a) they meet in the bottom right hand corner; in Figure 11.6(b) they meet in the top right hand corner.

So, since the slope of the second line, $\dfrac{-\alpha}{2}$, is determined by α, then a small change in α leads to a small change in the slope, and because the lines are nearly parallel then this leads to a disproportionately large change in point of intersection, or solution of the equations.

This is what we call *ill-conditioning*. You might think it is something of a rarity. While it is true that it does not occur very frequently, it can appear in engineering situations often enough that it is useful that you are aware of the possibility of extreme sensitivity to rounding errors in certain circumstances.

DO THIS NOW! 11.6

A set of simultaneous equations are defined in matrix form as follows:

$$\begin{pmatrix} \beta & 1 \\ 4 & 2 \end{pmatrix} \begin{pmatrix} x \\ y \end{pmatrix} = \begin{pmatrix} 1 \\ 3 \end{pmatrix}$$

where β is a constant.

(a) If $\beta = 2$, it is not possible to find a solution. Explain in your own way why not, both algebraically and graphically.

(b) Solve the equations for the two cases where $\beta = 2.1$ and where $\beta = 1.9$. Explain in your own way both graphically and in algebraic terms how these two solutions illustrate the concept of ill-conditioning.

11.9 SOLVING THE "REAL" PROBLEMS

In Section 11.2, we introduced two small examples of where simultaneous equations arise in engineering. You now have a whole range of ways of dealing with two simultaneous equations in two unknowns. You should understand the process geometrically as well as algebraically. You should have sorted out how to use technology to solve equations and you should have noticed that technology can deal with larger sets of equations.

DO THIS NOW! 11.7

Finish Examples 11.2.1 and 11.2.2 in Section 11.2, by solving the equations by any means. In Example 11.2.1, take $R_1 = 1000$, $R_2 = 10000$, $R_3 = 5000$, $V = 1$.

Satisfy yourself that your answers are correct by using another solution method.

END OF CHAPTER 11 - MIXED EXERCISES – DO THESE NOW!

1. Solve the equations $\begin{matrix} 4x + 3y = 8 \\ 5x + 7y = 4 \end{matrix}$ using each of the methods (a) – (e) below. In each case list the advantages and disadvantages of the method as you see them.

   (a) Graphical solution methods.

   (b) Use of tabulation and zooming.

   (c) Algebraic direct elimination.

   (d) Cramer's Rule.

   (e) Use of technology.

2. Solve the following two sets of simultaneous equations using at least two independent ways to confirm your results.

   (a) $\begin{matrix} 4x + 5y = 7.89 \\ x - y = 1 \end{matrix}$ (b) $\begin{matrix} 1.23x + 4.56y = 7.89 \\ -0.12x - 3.45y = -6.78 \end{matrix}$.

3. The perimeter of a rectangular plate is 160 cm. If the length was 3% longer and the breadth 5% longer, the perimeter would be 166 cm. Find the length and breadth of the plate. (Hint: Let x = length and y = breadth.)

4. Using 9 machines of type A and 7 of type B, each making the same unit, a manufacturer's output is 287 units per day. If they use 7 of type A and 9 of type B, the output is 305 a day. How many are made per day on each type of machine?

5. Solve the general equations for x_1 and x_2 as generated in section 11.5.

$$a_{11}x_1 + a_{12}x_2 = b_1$$
$$a_{21}x_1 + a_{22}x_2 = b_2$$

You might solve these algebraically, or using a CAS system, or preferably both. Check your answer by seeing if it agrees with what was quoted in section 11.5.

6. Note this example generates 3 equations in 3 unknowns. Think about how you will handle this: perhaps use technology or wait until Chapter 13.

The vertical profile curve of a roller coaster is shown in Figure 11.7. To estimate the forces on such a ride it is necessary to work out the equation of this curve. The co-ordinates of three points on the profile may be seen in the frames in Figure 11.7.

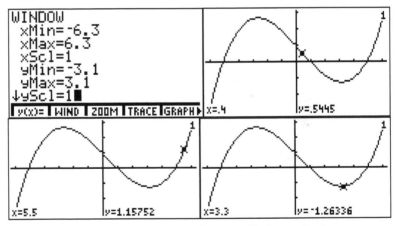

Figure 11.7 Roller Coaster profiles for question 6.

The profile over the range $0 < x < 6$ is to be modelled by the equation:

$w(x) = ax^2 + bx + c$.

Use the data in Figures 11.7 to write down three simultaneous equations for a, b, and c, and solve those equations to find a, b, and c. Check your answer by plotting the curve and seeing whether it does indeed provide the right shape.

12

Matrices

"Why," said the Dodo, "the best way to explain it is to do it." Lewis Carroll

12.1 MATRICES: WHAT ARE THEY, AND WHY DO YOU NEED THEM?

If you have read Chapter 11, you will have seen the use of what we called the "matrix form" of the simultaneous equations. This was recognition that the "shape" of the equations was square or rectangular, and that the position of different numbers or symbols within the rectangle had meaning. This is true in other systems too, engineering and otherwise (see later in this chapter). It leads us to define an object called a *matrix* (plural *matrices*) and to make rules for handling these matrices.

A **matrix** is *a rectangular array of numbers or symbols*. In this book we will denote matrices by a bold upper case letter.

Examples 12.1.1

All of the following are matrices:

$$A = \begin{pmatrix} 1 & 2 \\ 3 & 4 \end{pmatrix} \qquad B = \begin{pmatrix} 3 \\ 5 \end{pmatrix} \qquad X = \begin{pmatrix} x_1 \\ x_2 \end{pmatrix} \qquad D = \begin{pmatrix} 1 & 5 & -2 \end{pmatrix}$$

$$E = \begin{pmatrix} 0 & -1 \\ 2 & 4 \end{pmatrix} \qquad F = \begin{pmatrix} 8 \end{pmatrix}.$$

But the array $\begin{pmatrix} 1 & 2 \\ 3 & \end{pmatrix}$ is not a matrix as it is not rectangular.

When a matrix consists of a single column or row, it is often called a column or row vector, and some technologies talk of vectors of vectors. See Chapter 14.

The *order* of a matrix describes its shape, in the particular form:

(no. of rows) × (no. of columns). Read this as "number of rows by number of columns". The individual components of a matrix are called its *elements*.

In the matrices of Example 12.1.1, **A** is of order 2 × 2 ("two by two", a *square* matrix), **B** and **X** are 2 × 1, **D** is 1 × 3, **E** is 2 × 2 and **F** is 1 × 1.

Example 12.1.2 Matrix form of simultaneous equations

The simultaneous equations from Example 11.3.1

$$4x + 2y = 8$$

$$8x - 5y = -2$$

can be written in matrix form as follows:

$$\begin{pmatrix} 4 & 2 \\ 8 & -5 \end{pmatrix} \begin{pmatrix} x \\ y \end{pmatrix} = \begin{pmatrix} 8 \\ -2 \end{pmatrix}.$$

This involves one 2×2 matrix and two 2×1 matrices. We shall see how matrix operations help you to understand and solve these equations in section 12.4.

Example 12.1.3 Robots, computer games and computer graphics

Matrices play a crucial part in providing a mathematical description of the geometry needed for programming computer games, robots and graphics on a computer or video screen. We shall say more about this in section 12.5.

Here is a reminder again that as you work through this chapter you should remember to keep on **doing the exercises and activities** as you go.

12.2 ARITHMETIC AND ALGEBRAIC OPERATIONS WITH MATRICES

In this section you will meet the rules for handling matrices. Some of these rules will seem obvious. Some of them will seem odd and definitely not obvious, but they have been made the way they have for good reasons, which will become clear as you proceed.

12.2.1 Equality of matrices

Two matrices can only be equal if they are of the same order or shape, and *all* individual corresponding elements equal.

Example 12.2.1

If $\begin{pmatrix} x_1 \\ x_2 \end{pmatrix} = \begin{pmatrix} 1 \\ -3 \end{pmatrix}$ then $x_1 = 1$ and $x_2 = -3$.

12.2.2 Addition and subtraction of matrices

Adding and subtracting matrices is defined in an obvious way. That is, two matrices can only be added and subtracted if they are of the same order or shape. Then corresponding elements are added or subtracted.

Example 12.2.2

Using the matrices of Example 12.1.1:

$$A+E = \begin{pmatrix} 1 & 2 \\ 3 & 4 \end{pmatrix} + \begin{pmatrix} 0 & -1 \\ 2 & 4 \end{pmatrix} = \begin{pmatrix} 1+0 & 2+(-1) \\ 3+2 & 4+4 \end{pmatrix} = \begin{pmatrix} 1 & 1 \\ 5 & 8 \end{pmatrix}.$$

$$A-E = \begin{pmatrix} 1 & 2 \\ 3 & 4 \end{pmatrix} - \begin{pmatrix} 0 & -1 \\ 2 & 4 \end{pmatrix} = \begin{pmatrix} 1-0 & 2-(-1) \\ 3-2 & 4-4 \end{pmatrix} = \begin{pmatrix} 1 & 3 \\ 1 & 0 \end{pmatrix}$$ (Watch your signs closely!)

$$A \pm B = \begin{pmatrix} 1 & 2 \\ 3 & 4 \end{pmatrix} \pm \begin{pmatrix} 3 \\ 5 \end{pmatrix}$$ cannot be done as **A** and **B** are of different orders.

12.2.3 Scalar Multiplication

Scalar multiplication is the multiplication of a matrix by a scalar (or number), using the simple rule that all elements of the matrix are multiplied by the scalar.

Example 12.2.3

$$5A = 5 \begin{pmatrix} 1 & 2 \\ 3 & 4 \end{pmatrix} = \begin{pmatrix} 5\times 1 & 5\times 2 \\ 5\times 3 & 5\times 4 \end{pmatrix} = \begin{pmatrix} 5 & 10 \\ 15 & 20 \end{pmatrix}.$$

12.2.4 Matrix Multiplication

Matrix multiplication, or the multiplying together of two matrices, is sometimes but not always possible. It is defined in a strange and non-obvious way, but you will have the chance to see the sense of the definition when you see matrices being used in solving simultaneous equations, or in computer graphics.

To help you to see how it works, we shall first describe the process step by step, then do an example, and then discuss the theoretical points arising.

The method is so-called "row-by-column" multiplication. Here are the steps.

1. Take a row, say row i, from the first matrix, and a column, say column j, from the second matrix.

2. Lay these alongside each other.

3. Multiply corresponding elements, and add up the results of these multiplications.

4. The result of this goes into row i, column j of the product matrix.

5. Repeat for every possible combination of rows from first matrix and columns from second matrix, and this will give the product.

Example 12.2.4

In this example we work out **A.E** (with **A** and **E** as defined in Example 12.1.1).

$$A.E = \begin{pmatrix} 1 & 2 \\ 3 & 4 \end{pmatrix} \begin{pmatrix} 0 & -1 \\ 2 & 4 \end{pmatrix}.$$

Start with row 1, column 1.

Step 1 Row 1 from **A** is $\begin{pmatrix} 1 & 2 \end{pmatrix}$ and column 1 from **B** is $\begin{pmatrix} 0 \\ 2 \end{pmatrix}$.

Step 2 Lay these alongside each other (you have to slide the **A** row "round the corner"

$$\begin{pmatrix} 1 \\ 2 \end{pmatrix}\begin{pmatrix} 0 \\ 2 \end{pmatrix}.$$

Step 3 Multiply corresponding elements and add the results: $1\times 0 + 2 \times 2 = 4$.

Step 4 The value 4 goes into row 1, column 1 of the product matrix.

Step 5 is to repeat for every possible combination of rows and columns. Thus

Row1, Column 2 gives $1\times(-1)+2\times 4 = 7$,

Row2, Column 1 gives $3\times(0)+4\times 2 = 8$,

Row2, Column 2 gives $3\times(-1)+4\times 4 = 13$.

This has dealt with all possible combinations, so the product is

$$\mathbf{A.E} = \begin{pmatrix} 4 & 7 \\ 8 & 13 \end{pmatrix}.$$

This process will soon become familiar as you practice, although like all of us you will be very vulnerable to mistakes such as getting signs wrong!

There are some general points to make here.

You cannot just multiply any old pair of matrices together – they must clearly have the right shapes – we say they must be *compatible*. This is because in Step 2, when we lay the row from the first matrix alongside the column from the second, they must "match up". That is, there must be something for each element to multiply! For this to be true, the number of columns in the first matrix must match the number of rows in the second.

In our example **A.E** is order (2×2) times order (2×2). A shorthand check is that the middle two numbers when you write the orders down must be equal for the multiplication to be possible. Then the outside numbers give the shape or order of the result. In our example the middle numbers are both 2 and the result is indeed (2×2).

Here are some other examples.

Example 12.2.5

Note that $\mathbf{A.B} = \begin{pmatrix} 1 & 2 \\ 3 & 4 \end{pmatrix}\begin{pmatrix} 3 \\ 5 \end{pmatrix}$ is (2×2) times (2×1). The middle numbers match (both 2) and the result is predicted by the outside numbers to be (2×1).

The result (you check this!) is $\mathbf{A.B} = \begin{pmatrix} 13 \\ 29 \end{pmatrix}$.

But $\mathbf{B.A} = \begin{pmatrix} 3 \\ 5 \end{pmatrix}\begin{pmatrix} 1 & 2 \\ 3 & 4 \end{pmatrix}$ is (2×1) times (2×2). The middle numbers 1 and 2 do not match and so the matrix multiplication simply cannot be done.

You may have noticed that in the last example $\mathbf{A.B}$ and $\mathbf{B.A}$ were not the same – indeed one of them cannot be done! It is true that even in cases where you can reverse the order of multiplication and still do the operation, the result is not necessarily the same. We say that matrix multiplication is *non-commutative*. You therefore have to be careful, when writing down operations involving matrix multiplication, to retain the order in which multiplications occur. Pre-multiplying (multiplying in front) and post-multiplying (multiplying behind) must be retained.

Example 12.2.6

$\mathbf{A.E} = \begin{pmatrix} 1 & 2 \\ 3 & 4 \end{pmatrix}\begin{pmatrix} 0 & -1 \\ 2 & 4 \end{pmatrix} = \begin{pmatrix} 4 & 7 \\ 8 & 13 \end{pmatrix}$ from above,

but (check our working here!)

$\mathbf{E.A} = \begin{pmatrix} 0 & -1 \\ 2 & 4 \end{pmatrix}\begin{pmatrix} 1 & 2 \\ 3 & 4 \end{pmatrix} = \begin{pmatrix} -3 & -4 \\ 14 & 20 \end{pmatrix}$ which is not the same.

DO THIS NOW! 12.1

If you have not already done so, check the working in Examples 12.2.5 and 12.2.6.

12.2.5 Matrix Division

This is a quick and easy one to deal with: YOU CANNOT DIVIDE MATRICES. There is no matrix division defined! That was easy. On to the next thing then.

12.2.6 Transpose of a matrix

A final operation which may be needed and which does not correspond to a standard arithmetic operation, is finding the ***transpose*** of a matrix. This simply means swapping rows and columns over, so Row 1 becomes Column 1, etc. The notation used is either $\mathbf{A}^T$ or $\mathbf{A}'$.

Example 12.2.7

$\mathbf{A} = \begin{pmatrix} 1 & 2 \\ 3 & 4 \end{pmatrix}$ so $\mathbf{A}^T = \begin{pmatrix} 1 & 3 \\ 2 & 4 \end{pmatrix}$.

$\mathbf{B} = \begin{pmatrix} 3 \\ 5 \end{pmatrix}$ so $\mathbf{B}^T = \begin{pmatrix} 3 & 5 \end{pmatrix}$.

12.3 MATRICES AND TECHNOLOGY

Your various technologies will have ways of entering matrices, and performing the operations you have just met with them.

DO THIS NOW! 12.2

Find out how your favourite technology handles matrices. Can it deal with all we have defined above? Can it handle both algebraic and number-only matrices? Get some practice by doing ALL these problems.

You are given matrices $\mathbf{G} = \begin{pmatrix} 4 & -1 \\ 2 & 6 \end{pmatrix}$ $\mathbf{H} = \begin{pmatrix} 0 & 3 \\ -1 & 4 \end{pmatrix}$ $\mathbf{K} = \begin{pmatrix} 5 \\ 2 \end{pmatrix}$.

Calculate the following expressions, where possible, by hand and also checking your results with your favourite technology. If an expression is not possible, explain why.

$2\mathbf{G}$ $3\mathbf{G} + \mathbf{H}$ $\mathbf{G} + \mathbf{K}$ $\mathbf{G} - \mathbf{H}$ $\mathbf{G}.\mathbf{H}$ $\mathbf{H}.\mathbf{G}$ $\mathbf{G}.\mathbf{K}$ $\mathbf{K}.\mathbf{G}$

$\mathbf{G}^2$ $\mathbf{H}^2$ $\mathbf{K}^2$ $\mathbf{G}^T$ $\mathbf{K}^T$ $\mathbf{K}^T.\mathbf{K}$ $\mathbf{K}.\mathbf{K}^T$

12.4 MATRICES AND SIMULTANEOUS EQUATIONS – THE MATRIX INVERSE

You have seen in Chapter 11 how the mathematics of simultaneous equations works. Matrix theory provides a strong framework within which to understand linear simultaneous equations. First you will need two more definitions.

12.4.1 The determinant of a square matrix

You have already met the idea of a determinant. This can now be linked to matrix theory as follows. We can define the *determinant* of a square 2×2 matrix $\mathbf{A} = \begin{pmatrix} a_{11} & a_{12} \\ a_{21} & a_{22} \end{pmatrix}$ as

$$\det(\mathbf{A}) = |\mathbf{A}| = \begin{vmatrix} a_{11} & a_{12} \\ a_{21} & a_{22} \end{vmatrix} = a_{11} \times a_{22} - a_{21} \times a_{12} \ ,$$

just as we defined a determinant in Chapter 11. Note this is just a *number*, associated with a square matrix, and that the brackets round the matrix have turned into straight lines in the determinant. The determinant can be defined for larger square matrices too, but we shall postpone that discussion for now.

12.4.2 The identity matrix

You will also need the idea of the *identity matrix* **I**. This is the matrix that does the job of the number 1, in that multiplying by it changes nothing. The identity matrix is square and there is one for each possible size.

The 2×2 identity is $\mathbf{I} = \begin{pmatrix} 1 & 0 \\ 0 & 1 \end{pmatrix}$. The 3×3 identity is $\mathbf{I} = \begin{pmatrix} 1 & 0 & 0 \\ 0 & 1 & 0 \\ 0 & 0 & 1 \end{pmatrix}$. The general

pattern is to have a 1 all the way down the diagonal from top left to bottom right (the so-called "main diagonal") and zeroes everywhere else.

Your technology should find determinant and identity; see for example Figure 12.1.

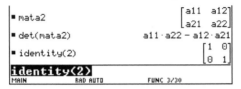

Figure 12.1 Matrix operations in a CAS system (TI-89).

DO THIS NOW! 12.3

(a) Find out NOW how *your* technology handles determinants and identities.

(b) Calculate **I.A** and **A.I** and check that the answer in each case is just **A**.

12.4.3   The idea of a matrix inverse

Finally we come to how these ideas help with simultaneous equations. First have a look at a single equation in the unknown x:

$$a.x = b \ .$$

If you wish to solve this, *without* using division (matrices have no division!) you cannot "divide both sides by a". What you must do is find a number which when you *multiply a by it* then you get 1. We call this the inverse or reciprocal of a, and usually write it a^{-1}. Then you can multiply both sides by this inverse and solve the equation.

$$a^{-1}.a.x = a^{-1}.b \text{ and so } 1..x = a^{-1}.b \text{ which is just } x = a^{-1}.b \ .$$

Example 12.4.1

If $2.x = 4$, then $a = 2$ and $a^{-1} = 0.5$, so we can solve this equation without division by multiplying both sides by 0.5.

$$0.5 \times 2.x = 0.5 \times 4 \text{ and so } x = 2 \ .$$

This idea carries over into simultaneous equations, if you write the equations in matrix form. Recall that the equations in general form may be written

$$a_{11}x_1 + a_{12}x_2 = b_1$$
$$a_{21}x_1 + a_{22}x_2 = b_2$$

and that these can be written in matrix form as

$$\begin{pmatrix} a_{11} & a_{12} \\ a_{21} & a_{22} \end{pmatrix} \begin{pmatrix} x_1 \\ x_2 \end{pmatrix} = \begin{pmatrix} b_1 \\ b_2 \end{pmatrix}.$$

Note now that, where the two matrices are written next to each other, that really does mean they are multiplied.

DO THIS NOW! 12.4

Satisfy yourself that if you multiply out the LHS matrices you really do get back to the original equations.

Now let's start using the power of the matrix notation. These equations can be written in shorthand form as $\mathbf{A.X = B}$,

where $\mathbf{A} = \begin{pmatrix} a_{11} & a_{12} \\ a_{21} & a_{22} \end{pmatrix}$, $\mathbf{X} = \begin{pmatrix} x_1 \\ x_2 \end{pmatrix}$, and $\mathbf{B} = \begin{pmatrix} b_1 \\ b_2 \end{pmatrix}$.

Now IF we can find an inverse $\mathbf{A}^{-1}$ which does actually satisfy $\mathbf{A}^{-1}.\mathbf{A} = \mathbf{I}$, then we can just pre-multiply both sides of this equation by the inverse, like this:

$$\mathbf{A}^{-1}.\mathbf{A.X} = \mathbf{A}^{-1}.\mathbf{B}$$

or
$$\mathbf{I.X} = \mathbf{A}^{-1}.\mathbf{B}$$

or
$$\mathbf{X} = \mathbf{A}^{-1}.\mathbf{B}$$

since multiplying by $\mathbf{I}$ is just like multiplying by 1.

Thus you can see that to get the solution of the simultaneous equations, all you have to do is calculate the matrix product $\mathbf{A}^{-1}.\mathbf{B}$. We return then to the matter of how you find this matrix inverse, if you can!

12.4.4 Finding the matrix inverse

It turns out that for the 2×2 matrix $\mathbf{A} = \begin{pmatrix} a_{11} & a_{12} \\ a_{21} & a_{22} \end{pmatrix}$, the inverse is given by

swapping the elements on the main diagonal (top left to bottom right), changing the

signs on the other diagonal, and multiplying the matrix by $\dfrac{1}{\det(\mathbf{A})}$.

That is:
$$\mathbf{A}^{-1} = \frac{1}{\det(\mathbf{A})} \begin{pmatrix} a_{22} & -a_{12} \\ -a_{21} & a_{11} \end{pmatrix} .$$

For larger square matrices the calculation is more complicated but we delay that discussion until chapter 13.

Example 12.4.2

If $\mathbf{G} = \begin{pmatrix} 4 & -1 \\ 2 & 6 \end{pmatrix}$ then

$$\mathbf{G}^{-1} = \frac{1}{\det(\mathbf{G})} \begin{pmatrix} 6 & -(-1) \\ -(2) & 4 \end{pmatrix} = \frac{1}{26} \begin{pmatrix} 6 & 1 \\ -2 & 4 \end{pmatrix} .$$

DO THIS NOW! 12.5

Find out NOW how your technology handles matrix inverses. (One example is shown in Figure 12.2.)

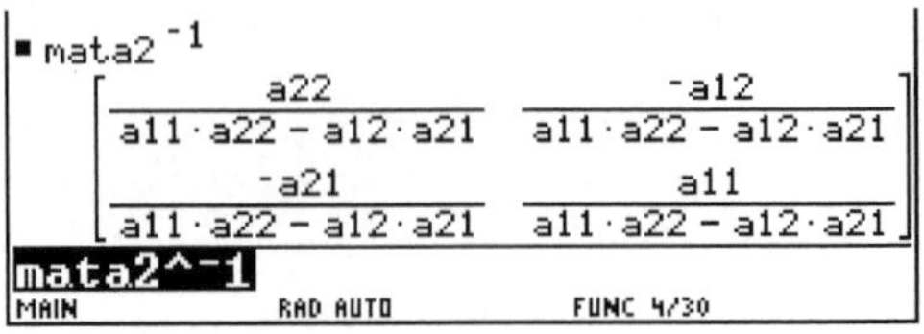

Figure 12.2 Finding a matrix inverse on CAS.

Find the inverse of $\mathbf{H} = \begin{pmatrix} 0 & 3 \\ -1 & 4 \end{pmatrix}$ by hand, and check that it works by calculating

both $\mathbf{H}^{-1}.\mathbf{H}$ and $\mathbf{H}.\mathbf{H}^{-1}$. Check your matrix inverse with your favourite technology.

Check (by hand and by machine) that for Example 12.4.2, $\mathbf{G}^{-1}.\mathbf{G}$ really does give

$\mathbf{I} = \begin{pmatrix} 1 & 0 \\ 0 & 1 \end{pmatrix}$. Incidentally, check that $\mathbf{G}.\mathbf{G}^{-1}$ also gives the same answer.

12.4.5 Using the matrix inverse to solve linear simultaneous equations

We can now apply these ideas to an example of simultaneous equations.

Example 12.4.3

Consider the equations

$$4x_1 - x_2 = 2$$
$$2x_1 + 6x_2 = 3$$

In matrix form these are

$$\begin{pmatrix} 4 & -1 \\ 2 & 6 \end{pmatrix}\begin{pmatrix} x_1 \\ x_2 \end{pmatrix} = \begin{pmatrix} 2 \\ 3 \end{pmatrix},$$

which is

$$\mathbf{G.X} = \begin{pmatrix} 2 \\ 3 \end{pmatrix},$$

where $\mathbf{G}$ is as above, for which we know that $\mathbf{G}^{-1} = \dfrac{1}{26}\begin{pmatrix} 6 & 1 \\ -2 & 4 \end{pmatrix}$.

$$\mathbf{X} = \mathbf{G}^{-1}.\begin{pmatrix} 2 \\ 3 \end{pmatrix}$$

The solution is then

$$= \frac{1}{26}\begin{pmatrix} 6 & 1 \\ -2 & 4 \end{pmatrix}\begin{pmatrix} 2 \\ 3 \end{pmatrix}$$

$$= \frac{1}{26}\begin{pmatrix} 15 \\ 8 \end{pmatrix} = \begin{pmatrix} 0.577 \\ 0.308 \end{pmatrix}$$

Finally then

$$x_1 = 0.577 \text{ and } x_2 = 0.308 \text{ to 3 d.p.}$$

DO THIS NOW! 12.6

Solve the simultaneous equations below by using the matrix inverse of the LHS matrix. In each case find the inverse by hand and check using technology. Check your answers by using a simultaneous equation solver in your technology, or using what you learnt in Chapter 11, or both!

(a) $\begin{aligned} 4x+3y &= 8 \\ 5x+7y &= 4 \end{aligned}$ (b) $\begin{aligned} 4x+5y &= 7.89 \\ x-\ y &= 1 \end{aligned}$ (c) $\begin{aligned} 1.23x+4.56y &=\ 7.89 \\ -0.12x-3.45y &= -6.78 \end{aligned}$.

12.4.6 Singular matrices

You might recall (Section 11.7) that there was a problem with simultaneous equations, which is identified if the determinant of the LHS is 0. The equations are then called singular. If this is the case, it means that there is not a unique solution to the equations – there are either no solutions at all, or an infinite number of solutions. If the matrix inverse is to be useful it must reflect this same fact.

To see how this shows itself, look back at how to calculate the inverse.

For the general 2×2 matrix $\mathbf{A} = \begin{pmatrix} a_{11} & a_{12} \\ a_{21} & a_{22} \end{pmatrix}$, then $\mathbf{A}^{-1} = \dfrac{1}{\det(\mathbf{A})}\begin{pmatrix} a_{22} & -a_{12} \\ -a_{21} & a_{11} \end{pmatrix}$.

You can see that to find the inverse you need to divide by the determinant of the matrix $\mathbf{A}$. If this is zero we cannot divide by it, so we simply cannot find the inverse, and so as before we cannot solve the equations uniquely. If $\det(\mathbf{A}) = 0$, then we call the matrix $\mathbf{A}$ *singular*. Correspondingly if $\det(\mathbf{A}) \neq 0$, then we call the matrix $\mathbf{A}$ *non-singular*. This is one of many strange uses of language that you should be getting used to in mathematics!

12.5 MATRICES AND GEOMETRICAL TRANSFORMATIONS

In this section we give you the opportunity to explore how matrices can be useful in changing or transforming 2-dimensional pictures – an essential part, for example, of computer graphics, computer games and, when expanded to three dimensions, mathematical description of the position of robot arms.

We start by setting up an activity for you to work through and describe what you find, and then we give a brief summary at the end of the section of what we think you should have found.

Begin by giving the matrix

$$\mathbf{P} = \begin{bmatrix} x_1 & x_2 & \cdot & \cdot & \cdot & \cdot \\ y_1 & y_2 & \cdot & \cdot & \cdot & \cdot \end{bmatrix}$$

geometrical meaning, by interpreting the pairs of values (x_i, y_i) (i.e. each column of the matrix) as points in the Cartesian (x-y) plane.

As an example we shall use the matrix which we shall give the name **SHAPE**

$$\mathbf{SHAPE} = \begin{bmatrix} 1 & 1 & 2 & 3 & 3 & 1 \\ 1 & 2 & 3 & 2 & 1 & 1 \end{bmatrix}.$$

Plot this set of points on the x-y plane, in the order they occur from left to right in the matrix, joining them in that order. This gives the diagram shown in Figure 12.3.

	Matrix "shape"					
x	1	1	2	3	3	1
y	1	2	3	2	1	1

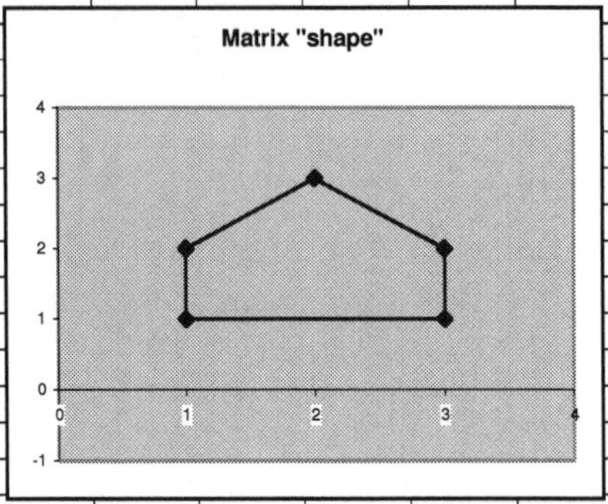

Figure 12.3 Spreadsheet extract representing matrix "**SHAPE**".

We have produced this diagram in a spreadsheet (Microsoft Excel actually), and have also carried out all the other calculations here in that medium too. We recommend strongly that the spreadsheet is a tool you should learn to use, both in designing your own operations and using built-in functions. See the end of chapter exercises for a little more information. Watch out for well-hidden tricks though. For instance if you want to use matrix commands such as MMULT in Excel, you must hold down both Shift and Control keys when pressing Enter to carry out the operation, but this is not well publicised!

DO THIS NOW! 12.7

For each of the matrices **T** below in turn, form the product **T** × **SHAPE** – that is, pre-multiply the matrix **SHAPE** by each matrix in turn.

The resulting matrix in each case should be the same order (shape!) as the matrix **SHAPE**, and so could be represented as a figure on the same diagram as the original matrix. Make a conjecture about the geometrical effect that multiplying by the matrix has had on the picture; then actually plot the new transformed matrix on your diagram and describe what you see, commenting on the effect the matrix **T** has had on your picture.

Repeat for all the matrices below. Can you draw any general conclusions about particular types of matrix? Try to do that and then refer to the end of this box to see what we suggest.

For example if $\mathbf{T} = \begin{bmatrix} 1 & 0 \\ 0 & -1 \end{bmatrix}$ then $\mathbf{T} \times \mathbf{SHAPE} = \begin{bmatrix} 1 & 1 & 2 & 3 & 3 & 1 \\ -1 & -2 & -3 & -2 & -1 & -1 \end{bmatrix}$.

Plot this and you will see that the plotted object has been turned upside down – or more precisely reflected in the x-axis.

(DO THIS NOW 12.7! continued overleaf)

DO THIS NOW! 12.7 (Continued)

(i) $T = \begin{bmatrix} 0.9 & 0 \\ 0 & 0.9 \end{bmatrix}$ (ii) $T = \begin{bmatrix} 0.5 & 0 \\ 0 & 0.5 \end{bmatrix}$ (iii) $T = \begin{bmatrix} -0.5 & 0 \\ 0 & -0.5 \end{bmatrix}$

(iv) $T = \begin{bmatrix} 1 & 0 \\ 0 & -1 \end{bmatrix}$ (v) $T = \begin{bmatrix} 0.5 & 0 \\ 0 & -0.5 \end{bmatrix}$ (vi) $T = \begin{bmatrix} 0 & 1 \\ 1 & 0 \end{bmatrix}$

(vii) $T = \begin{bmatrix} 0 & 1 \\ -1 & 0 \end{bmatrix}$ (viii) $T = \begin{bmatrix} 0 & -1 \\ 1 & 0 \end{bmatrix}$ (ix) $T = \begin{bmatrix} \cos(\pi/3) & -\sin(\pi/3) \\ \sin(\pi/3) & \cos(\pi/3) \end{bmatrix}$

(x) $T = \begin{bmatrix} \cos(\pi/6) & -\sin(\pi/6) \\ \sin(\pi/6) & \cos(\pi/6) \end{bmatrix}$ (xi). $T = \begin{bmatrix} 0.5 & 0 \\ 0 & 0.5 \end{bmatrix}\begin{bmatrix} \cos(\pi/3) & -\sin(\pi/3) \\ \sin(\pi/3) & \cos(\pi/3) \end{bmatrix}$

Create your own matrices to explore all this further. Remember RADIANS!

You may have reached the following conclusions:

1. A matrix of the form $T = \begin{bmatrix} a & 0 \\ 0 & a \end{bmatrix}$ scales the picture by a factor of a, with the centre of the scaling at the origin. A more general result is that a matrix of the form $T = \begin{bmatrix} a & 0 \\ 0 & b \end{bmatrix}$ scales the picture by a factor of a in the x direction and b in the y direction.

2. A matrix of the form $T = \begin{bmatrix} -1 & 0 \\ 0 & 1 \end{bmatrix}$ reflects the picture in the x-axis. A matrix of the form $T = \begin{bmatrix} 1 & 0 \\ 0 & -1 \end{bmatrix}$ reflects the picture in the y-axis.

3. A matrix of the form $T = \begin{bmatrix} \cos(\theta) & -\sin(\theta) \\ \sin(\theta) & \cos(\theta) \end{bmatrix}$ rotates the picture through an angle θ in the anti-clockwise direction, with the origin as centre.

4. You can get more complex effects by combining the simple matrices together – but remember that matrix multiplication is not commutative, so the order of matrix multiplication matters. For example try the effect of matrix (vi) on **SHAPE** followed by matrix (vii) on the result of that, and see what happens; and then try them on **SHAPE** in reverse order. You will see the result in each case is quite different.

DO THIS NOW! 12.8

Any animation fans might like to look at designing a shape of their own, then designing matrices to make it spiral off into the distance. For instance you could create a matrix by combining say a rotation of $45°$ together with a shape reduction of 0.9, and then applying this matrix repeatedly and successively.

END OF CHAPTER 12 - MIXED EXERCISES – DO ALL THESE NOW!

1. You are given matrices **A** , **B** and **C** where

$$A = \begin{pmatrix} 1 & 3 \\ 2 & 6 \end{pmatrix} \qquad B = \begin{pmatrix} 3 & -1 \\ 7 & 6 \end{pmatrix} \qquad C = \begin{pmatrix} 1 \\ 2 \end{pmatrix}$$

Calculate the following expressions where possible, by hand and also checking your results with technology. If an expression is not possible, explain why.

2**A**	**A** + 4**B**	**A** + **C**	**A** – **B**	**A**×**B**	**B**×**A**
A×**C**	**C**×**A**	A^2	B^2	C^2	A^T
C^T	$C^T×C$	$C×C^T$	det(**A**)	det(**C**)	det(**B**)
A + 3**I**	**A** / **B**	C^{-1}	B^{-1}		

2. Where possible, and using the matrices **A**, **B**, and **C** as defined in question 1, with

$$X = \begin{pmatrix} x_1 \\ x_2 \end{pmatrix},$$ use the matrix inverse to solve the following matrix equations for x_1

and x_2:

    (a)   **B.X** = **C**

    (b)   **A.X** = **C**

Explain with the aid of diagrams or otherwise any difficulties you meet in trying to solve these two sets of equations.

13

More Linear Simultaneous Equations

"Mathematical skills are like any other kind.... If you are learning to play the piano, you usually start by practicing under supervision" Ralph Boas

13.1 LARGER SETS OF SIMULTANEOUS EQUATIONS

In chapters 11 and 12 you found out about how to handle two simultaneous equations in two unknowns, as well as meeting some matrix theory. The "paper and pencil" methods you met there do not extend easily to larger systems of equations. For instance finding bigger determinants, or inverses of bigger matrices, using theoretically based techniques, is clumsy and not practical, particularly when engineering systems can lead to sets of hundreds, thousands and even millions of equations. This is not uncommon, and you will meet, and deal with, an example of such a system in this chapter in section 13.5.

However, the theoretical principles you have met, such as the crucial role played by the determinant of the LHS of a system of equations, carry over. The only question is how to define and calculate matrix operations in such a way that the job is done. Briefly, we recommend that for systems of more than two equations, you apply appropriate technology to find determinants and matrix inverses, or to solve simultaneous equations. However we think it is useful for you to have some idea of the kind of methods your technology is using. This chapter is about these so-called numerical methods and some examples of using the technology to implement them.

Overall, there are two classes of systematic solution method for linear simultaneous equations suitable for use in technology: ***direct*** and ***iterative***. You will meet a direct method (an elimination method) in section 13.2, and some iterative methods in section 13.3. There are other methods too, but in this book we will restrict ourselves to just these.

Before going on, here is a reminder again that as you work through this chapter, just remember to keep on **doing the exercises and activities** as you go.

13.2 ELIMINATION METHODS: GAUSSIAN ELIMINATION

An elimination method is perhaps the most commonly used type of direct solution technique. Such a method appears often in the simultaneous equation solvers built into graphic calculators, albeit hidden behind a button or menu.

The general idea is to *systematically* eliminate all variables except one, solve for that one, and then *systematically* substitute backwards to find the others. You will notice that we emphasise the word *systematic*: this is because if a machine is to be programmed to implement a method, then you cannot rely on mathematical flair and spotting short cuts - the machine is stupid and will only do exactly what it is told, so the method must have no surprises! The first method we present here is *Gaussian elimination* and we do so by means of a simple 2 by 2 (two equations in two unknowns) example. After that we apply the method to a larger system.

Example 13.2.1: A 2 by 2 example of Gaussian Elimination

Solve for x_1 and x_2, by Gaussian elimination, the equations
$$\begin{array}{ll} 5x_1 + 4x_2 = 2 & (13.1) \\ 2x_1 - 4x_2 = 1 & (13.2) \end{array}.$$

The general strategy is to use equation (13.1) to eliminate x_1 from (13.2) by subtracting an appropriate multiple of (13.1) from (13.2). You then solve the resulting equation for x_2, and substitute back into (13.1) and solve for x_1.

Subtract $\frac{2}{5}$ (13.1) from (13.2) $\qquad\qquad -5.6x_2 = 0.2 \quad (13.2')$.

($\frac{2}{5}$ (13.1) *makes the* $5x_1$ *into* $2x_1$, *and therefore* x_1 *disappears as equations subtracted*)

Solve (13.2') for x_2 $\qquad\qquad x_2 = \dfrac{0.2}{-5.6} = -0.0357$.

Solve (13.1) for x_1 $\qquad\qquad x_1 = \dfrac{2 - 4x_2}{5} = 0.4286$.

The method can be generalised for any number of equations, eliminating each variable in turn. The systematic element comes from addressing the equations in a standard order. In the *elimination* phase for instance, you use equation (1) to eliminate x_1 from all equations below that, then the new equation (2) to eliminate x_2 from all equations below that, and so on. You then enter the *back-substitution* phase starting with the last variable, which should by then be alone in an equation, so you can find it. You go back one equation, and knowing the last variable you can find the last-but-one, and so on back to x_1. This method becomes progressively less efficient as the number of equations increases.

Perhaps a wordy explanation is not the best way, and you will be able to see better what is going on by looking at a 3 by 3 example.

Example 13.2.2: A 3 by 3 example of Gaussian Elimination

Solve these simultaneous equations by Gaussian elimination:

$$x_1 - 2x_2 + 3x_3 = 9 \quad (13.3)$$
$$- x_1 + 3x_2 \quad\quad = -4 \quad (13.4) \ .$$
$$2x_1 - 5x_2 + 5x_3 = 17 \quad (13.5)$$

The layout of the equations helps to keep track. Write these equations as an *augmented matrix* - that is, the LHS matrix next to the RHS. Thus anything you do to a row is like doing that thing to both sides of an equation. Each column represents the coefficients of a variable, x_1 in column 1 etc. Here is the augmented matrix:

$$\begin{bmatrix} 1 & -2 & 3 & 9 \\ -1 & 3 & 0 & -4 \\ 2 & -5 & 5 & 17 \end{bmatrix} .$$

Now we set the elimination stage of the algorithm going:

Step 1

Use Row1 (R1) to eliminate x_1 from Row 2 (R2) and Row 3 (R3) as follows:

Do R2 = R2 - (-1/1)R1:

(Note R1 is multiplied by $\left(\dfrac{\text{coeff. of } x_1 \text{ in } (2)}{\text{coeff. of } x_1 \text{ in } (1)} \right)$ to make sure x_1 disappears from R2.)

Do R3 = R3 - (2/1)R1:

(This time the factor is (2/1) to make sure x_1 disappears from R3)

The augmented matrix – which represents the equations – thus becomes:

$$\begin{bmatrix} 1 & -2 & 3 & 9 \\ 0 & 1 & 3 & 5 \\ 0 & -1 & -1 & -1 \end{bmatrix} .$$

The intended effect is that column 1 has a 1 in the x_1 position with zeroes below that. We are left with R2 and R3 involving x_2 and x_3 only.

Step 2

Use the new R2 to eliminate x_2 from R3 as follows:

Do R3 = R3 - (-1/1)R2:   Factor this time is $\left(\dfrac{\text{coeff. of } x_2 \text{ in } (3)}{\text{coeff. of } x_2 \text{ in } (2)} \right)$.

This gives $\begin{bmatrix} 1 & -2 & 3 & 9 \\ 0 & 1 & 3 & 5 \\ 0 & 0 & 2 & 4 \end{bmatrix} .$

Step 3

Make the x_1 coefficient in R1, the x_2 coefficient in R2 and the x_3 coefficient in R3 equal to 1 (if they are not already so) by dividing through each row (equation) by the appropriate number. This gives:

$$\begin{bmatrix} 1 & -2 & 3 & 9 \\ 0 & 1 & 3 & 5 \\ 0 & 0 & 1 & 2 \end{bmatrix}.$$

This last matrix is in so-called *row echelon form.*

The lower left triangle is all 0s and the *main diagonal* (top left diagonally downwards) is all 1s. The first three columns, representing the LHS, have become an *upper triangular* matrix. Doing things in the order we have makes sure that later steps do not mess up what was done in the earlier steps. We now translate this latest augmented matrix back into the set of equations that we are left with:

$$x_1 - 2x_2 + 3x_3 = 9 \qquad (13.6)$$
$$x_2 + 3x_3 = 5 \qquad (13.7).$$
$$x_3 = 2 \qquad (13.8)$$

You can see that the row echelon form that we have produced means that:

- it is easy to solve the third equation (13.8) for x_3: $x_3 = 2$;

- back-substituting this value, it is easy to solve the second equation (13.7) for x_2:

$$x_2 + 3 \times 2 = 5$$

$$\text{or } x_2 = 5 - 6 = -1 \text{ ;}$$

- finally, back-substituting again, it is easy to solve the first equation (13.6) for x_1:

$$x_1 - 2 \times (-1) + 3 \times 2 = 9$$

$$\text{or } x_1 = 9 - 6 - 2 = 1 \text{ .}$$

At the end, you should check that your answer is right by two means, as in **DO THIS NOW! 13.1.**

DO THIS NOW! 13.1

Check that the answers in Example 13.2.2 are right by two means:-

1. Substitute the values obtained for the unknowns back into the original equations to see if they work and really do make the LHS equal to the RHS.

2. Solve the original equations on your favourite piece of technology, which should handle three equations just as easily as it handles two!

DO THIS NOW! 13.2

Repeat the whole process with the equations below, checking your own answers as in DO THIS NOW 13.1!:

$$5x_1 - 4x_2 + 2x_3 = 9$$
$$x_1 - 3x_2 \qquad = -2 \; .$$
$$3x_1 + 4x_2 + 2x_3 = 14$$

DO THIS NOW! 13.3

Some technology has these processes built in, for instance as shown in the graphic calculator screens in Figure 13.1. Find out NOW whether the technology available to you has such features built in, perhaps under a name such as "row operations".

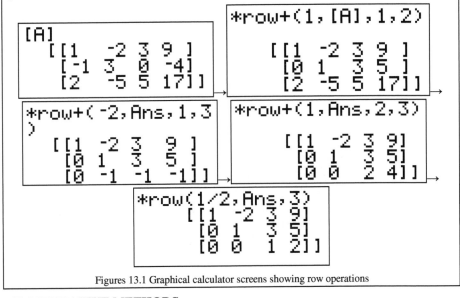

Figures 13.1 Graphical calculator screens showing row operations

13.3 ITERATIVE METHODS
13.3.1 What is an iterative method?

An iterative approach involves making a guess at the solution of the equations, and then using a systematic algorithm (sequence of calculating rules) to progressively and repeatedly update this guess so that it converges to (or approaches closer and closer to) the solution of the equations. There is no guarantee in every case that such a guess-and-correct approach will converge towards the solution of the equations, and some theoretical basis will be needed.

There are several different iterative approaches to solving linear simultaneous equations. You will meet here only two of them – the Jacobi method and the Gauss-Seidel method – but you should be aware that the use of numerical methods is a highly developed area. These examples give you an idea of major features though.

13.3.2 The Jacobi method

The general idea of the Jacobi method is this:

- You algebraically solve "equation (1)" for x_1 in terms of the other variables, "equation (2)" for x_2, and so on.

- You guess starting values for x_1, x_2 and so on.

- You use your forms of the equations to find new values for x_1, x_2 and so on.

- You repeat the previous step until you are getting no further change (within your required accuracy) in x_1, x_2 and so on. As they are not changing any more, these values of x_1, x_2 etc. make the LHS equal to the RHS, and so they are the solutions to the equations.

Example 13.3.1 An example using the Jacobi method

Solve for x_1 and x_2 using the Jacobi method

$$8x_1 + x_2 = 2 \qquad (13.9)$$
$$2x_1 - 9x_2 = 1 \qquad (13.10)$$

Step 1   Solve (13.9) for x_1 and (13.10) for x_2.

$$x_1 = \frac{2 - x_2}{8} \qquad (13.11)$$
$$x_2 = \frac{1 - 2x_1}{-9} \qquad (13.12)$$

Note that (13.11) and (13.12) are just the original two equations rewritten.

Step 2   Make a guess at the solution, say $x_1 = 0$ and $x_2 = 0$.

Step 3   Use (13.11) and (13.12) repeatedly to calculate updated values of x_1 and x_2. (Don't use the new values until each complete cycle of updating is complete.) Keep going until the values of x_1 and x_2 are not changing any more to within the accuracy that you want to work. When that happens, then x_1 and x_2 are the solutions to the original equations.

Here are the figures which result:

Step	0	1	2	3	4
x_1	0	0.250	0.264	0.257	0.257
x_2	0	-0.111	-0.056	-0.054	-0.054

It seems that things stop changing at step 4, so the solution has converged (to three decimal places) to a solution $x_1 = 0.257$ and $x_2 = -0.054$.

DO THIS NOW! 13.4

See if you can reproduce (and thus check) our figures **NOW**. Check the solutions now in some independent ways – by substituting into the original equations, by Cramer's rule, by using your technology and so on.

Repeat the process for the equations

$$10x_1 + x_2 = 3 \qquad (13.13)$$
$$3x_1 - 8x_2 = 2 \qquad (13.14)$$

13.3.2 The Gauss-Seidel method

This is entirely similar to the Jacobi method, except that you use each new value as soon as it is calculated, instead of waiting for the end of each cycle. It is a common feature of numerical methods that such seemingly small changes can make a significant difference to the outcome. You might guess that this change is going to speed up the convergence, so follow through the calculation to see if that is the case.

Example 13.3.2

Here the equations of Example 13.3.1 are solved using the Gauss-Seidel method.

Step	0	1	2	3
x_1	0	0.25	0.257	0.257
x_2	0	-0.056	-0.054	-0.054

The process converges to the same answers as before but with one iteration fewer.

DO THIS NOW! 13.5

Repeat and check our workings **NOW**. You already know the solutions are correct!

Carry out both the Jacobi and Gauss-Seidel methods for equations (13.15) and (13.16). Comment on relative speed of convergence:
$$12x_1 + 2x_2 = 1 \quad (13.15)$$
$$x_1 - 9x_2 = 6 \quad (13.16)$$

Technical note *The convergence of these iterative procedures is* guaranteed *only if the LHS matrix is what is called* diagonally dominant: *the absolute value of the coefficient of the diagonal element for each equation* a_{ii} *is greater than the sum of the absolute values of the coefficients of all other elements in that row, that is*

$$\sum_{i=1}^{n} |a_{ij}| < |a_{ii}| \quad \begin{cases} i = 1,2,...., n \\ j \neq i \end{cases}.$$

This is often the case for "real" problems (as for instance in Section 13.6).

DO THIS NOW! 13.6

Extend the Jacobi and the Gauss-Seidel methods to three simultaneous equations, by solving the following to 4 decimal places and comparing how many iterations it takes in each case for the method to converge to the solution:

$$8x_1 - 2x_2 + x_3 = 9 \quad (13.17)$$
$$x_1 + 6x_2 - 2x_3 = -2 \quad (13.18)$$
$$-x_1 + 2x_2 - 8x_3 = 14 \quad (13.19)$$

You might find it best to use a spreadsheet to carry out your calculations. We do!

Check for diagonal dominance.

Check your own solution again by substituting answers you find back into the original equations, and also by using your technology.

13.4 THE GAUSS-JORDAN METHOD

We now return your attention to elimination methods, to see how all this relates to the matrix inverse. The Gaussian elimination method can be extended by a further, similar set of extra elimination steps, to solve the equations completely without requiring the back-substitution stages. Recall that you use "equation (1)" to eliminate x_1, "equation (2)" to eliminate x_2 and so on, until you have made the entire lower triangle full of zeroes, and the main diagonal holding nothing but 1s. Look back again at Example 13.2.2 if you want to remind yourself of these elimination steps.

With this "Gauss-Jordan" method, instead of back substituting, you carry on with the elimination process using the last equation to eliminate the last x from all preceding equations, and so on in reverse order all the way back to using "(2)" to eliminate x_2 from "(1)". You will see, in Section 13.5, that this incidentally gives a general way of calculating the inverse of a square matrix of any size.

Again, wordy descriptions are difficult to make clear, so here is an illustration of this process, continuing Example 13.2.2 from the end of the elimination phase.

Example 13.4.1 Gauss-Jordan elimination (continuing example 13.2.2)

Recall we reduced these equations (13.3), (13.4) and (13.5):

$$x_1 - 2x_2 + 3x_3 = 9 \quad (13.3)$$
$$-x_1 + 3x_2 \qquad = -4 \quad (13.4) \;.$$
$$2x_1 - 5x_2 + 5x_3 = 17 \quad (13.5)$$

to these (13.6), (13.7) and (13.8), using a series of systematic elimination steps:

$$x_1 - 2x_2 + 3x_3 = 9 \quad (13.6)$$
$$x_2 + 3x_3 = 5 \quad (13.7)$$
$$x_3 = 2 \quad (13.8)$$

or in terms of the augmented matrix:
$$\begin{bmatrix} 1 & -2 & 3 & 9 \\ 0 & 1 & 3 & 5 \\ 0 & 0 & 1 & 2 \end{bmatrix}.$$

Now instead of back-substitution, we continue the elimination process:

Step 4 Use R3 to eliminate x_3 from R1 and R2 by doing

R2 = R2 –(3/1) R3　　　and　　　R1 = R1 – (3/1) R3　　$\begin{bmatrix} 1 & -2 & 0 & 3 \\ 0 & 1 & 0 & -1 \\ 0 & 0 & 1 & 2 \end{bmatrix}.$

Step 5 Use R2 to eliminate x_2 from R1 by doing

R1 = R1 – (-2/1) R2　　　$\begin{bmatrix} 1 & 0 & 0 & 1 \\ 0 & 1 & 0 & -1 \\ 0 & 0 & 1 & 2 \end{bmatrix}.$

This last matrix is then in what we call *reduced row echelon form*, with 0s in the upper right triangle as well. Some technologies do this directly – see Figure 13.2.

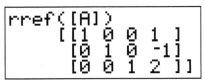

Figure 13.2 Reduced row echelon form on a graphic calculator.

In fact you can see that the first three columns have moved from being the LHS matrix to being an identity matrix. Converting the augmented matrix back into equations shows why this form is useful, because as you will see, they immediately give the solutions $x_1 = 1$, $x_2 = -1$ and $x_3 = 2$, as in the RH column.

DO THIS NOW! 13.7

Take the equations which you have solved by Gaussian elimination and back substitution in DO THIS NOW 13.2!

$$5x_1 - 4x_2 + 2x_3 = 9$$
$$x_1 - 3x_2 \qquad = -2$$
$$3x_1 + 4x_2 + 2x_3 = 14$$

and solve them by replacing the back substitution stages with the extra elimination steps of the Gauss-Jordan method. Verify that you get the same answers again!

13.5 FINDING A MATRIX INVERSE BY THE GAUSS-JORDAN METHOD

The Gauss-Jordan method gives a way of finding the inverse of any non-singular matrix, which is more practical and useful than other more theoretical methods. This is illustrated through Example 13.5.1.

Example 13.5.1

We shall use the LHS matrix of the equations of example 13.2.2 again:

$$x_1 - 2x_2 + 3x_3 = \ 9 \qquad (13.3)$$
$$- x_1 + 3x_2 \qquad = -4 \qquad (13.4) \ .$$
$$2x_1 - 5x_2 + 5x_3 = 17 \qquad (13.5)$$

Form the augmented matrix, but instead of putting in the RHS column, replace it by an identity matrix like this:

$$\begin{bmatrix} 1 & -2 & 3 & 1 & 0 & 0 \\ -1 & 3 & 0 & 0 & 1 & 0 \\ 2 & -5 & 5 & 0 & 0 & 1 \end{bmatrix} .$$

Then simply carry out the steps you would have carried out to perform Gauss-Jordan elimination, making Steps 1 – 5 so that at the end the first three columns are left forming an identity matrix. The last three columns will then have become the inverse of the original LHS matrix! The steps follow.

Step 1

Use Row1 (R1) to put zeroes into column 1 of Row 2 (R2) and Row 3 (R3).

R2 = R2 - (-1/1)R1.

R3 = R3 - (2/1)R1.

$$\begin{bmatrix} 1 & -2 & 3 & 1 & 0 & 0 \\ 0 & 1 & 3 & 1 & 1 & 0 \\ 0 & -1 & -1 & -2 & 0 & 1 \end{bmatrix}.$$

Step 2

Use the new R2 to put a zero into column 2 of R3 by doing:

R3 = R3 - (-1/1)R2 .

$$\begin{bmatrix} 1 & -2 & 3 & 1 & 0 & 0 \\ 0 & 1 & 3 & 1 & 1 & 0 \\ 0 & 0 & 2 & -1 & 1 & 1 \end{bmatrix}.$$

Step 3

Make the main diagonal coefficient in each row equal to 1 by dividing through each row by the appropriate number. This gives:

$$\begin{bmatrix} 1 & -2 & 3 & 1 & 0 & 0 \\ 0 & 1 & 3 & 1 & 1 & 0 \\ 0 & 0 & 1 & -0.5 & 0.5 & 0.5 \end{bmatrix}.$$

Step 4

Use R3 to create zeroes in column 3 of rows R1 and R2 by doing

R2 = R2 -3R3.

R1 = R1 - 3R3.

$$\begin{bmatrix} 1 & -2 & 0 & 2.5 & -1.5 & -1.5 \\ 0 & 1 & 0 & 2.5 & -0.5 & -1.5 \\ 0 & 0 & 1 & -0.5 & 0.5 & 0.5 \end{bmatrix}.$$

Step 5

Use R2 to put a zero into column 2 of R1 by doing

R1 = R1 – (-2)R2.

$$\begin{bmatrix} 1 & 0 & 0 & 7.5 & -2.5 & -4.5 \\ 0 & 1 & 0 & 2.5 & -0.5 & -1.5 \\ 0 & 0 & 1 & -0.5 & 0.5 & 0.5 \end{bmatrix}.$$

Now you can see that the first three columns have become an identity matrix, so the inverse of the original LHS matrix is the last three columns.

$$\begin{bmatrix} 7.5 & -2.5 & -4.5 \\ 2.5 & -0.5 & -1.5 \\ -0.5 & 0.5 & 0.5 \end{bmatrix} \qquad (13.20)$$

Using technology to check gives the right hand side of the matrix as

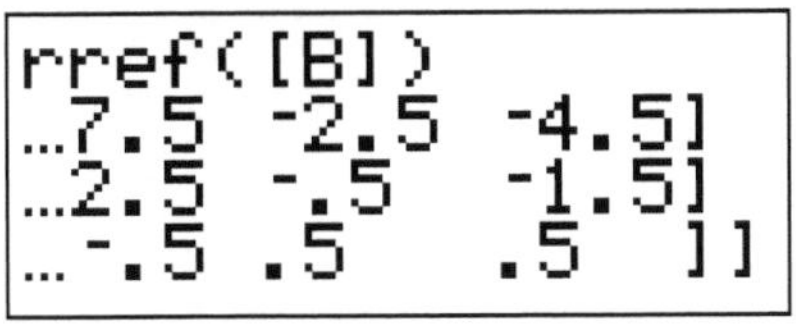

Figure 13.3 Graphic calculator display of Gauss-Jordan result.

DO THIS NOW! 13.8

Check that (13.20) really is the inverse matrix, first by multiplying it by the original

LHS matrix $\begin{pmatrix} 1 & -2 & 3 \\ -1 & 3 & 0 \\ 2 & -5 & 5 \end{pmatrix}$ and seeing if it satisfies $\mathbf{A}^{-1}.\mathbf{A} = \mathbf{A}.\mathbf{A}^{-1}=\mathbf{I}$, then by the

most convenient technology. Use the row operations of the Gauss-Jordan method to

find the inverse of the matrix $\begin{pmatrix} 5 & -4 & 2 \\ 1 & -3 & 0 \\ 3 & 4 & 2 \end{pmatrix}$, and check your answer as above.

(You might find a spreadsheet to be the best tool to help you here with the Gauss-Jordan calculations.)

13.6 ENGINEERING CASE STUDY: HEATING AND COOLING
13.6.1 The engineering problem

Simultaneous linear equations arise in many practical situations. One such is the use of so-called finite difference methods to model heat conduction situations. Similar methods are widely used to model for example electric, gravitational and magnetic fields, fluid flow, and stress and strain. We shall not do a full treatment here, but give a brief indication of where the equations come from and the issues that arise in solving them. One such issue is that you can very soon have a very large number of equations to solve.

In steady-state heat transfer problems in two dimensions, the temperature $T(x, y)$ at any point (x, y) satisfies an equation called the Laplace equation.

$$\frac{\partial^2 T}{\partial x^2} + \frac{\partial^2 T}{\partial y^2} = 0 \ .$$

This is a partial differential equation and most methods of solving it are beyond the scope of this book. However there are straightforward methods of getting an approximate solution to the equation by using the fact that it is a mathematical statement of the law of conservation of energy - in this case heat energy. One such method uses what are called finite differences. One way of implementing this method is to cover the region of interest with a square grid and to calculate only the approximate temperatures at the nodes of the grid. One such rectangular region, with temperatures specified around its boundaries, is shown covered with a coarse grid in Figure 13.4. We must find temperatures inside the region.

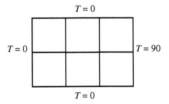

Figure 13.4 Finding the temperature in a rectangular region – figures are temperatures in degrees.

13.6.2 Extracting the equations from the physical situation

To see how the physical situation gives rise to mathematical equations, consider a little square area ABCD of the heat-conducting plate as shown in Figure 13.5.

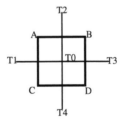

Figure 13.5 A magnified element of the region.

The values $T0$, $T1$, $T2$, $T3$, and $T4$ represent respectively the temperatures at the central point (or node) and the four surrounding equally distant nodes. These nodes correspond to places where the gridlines cross in Figure 13.4. For the sake of simplicity, suppose the length of side of the square ABCD is h, and that each of the temperature nodes $T1$, $T2$, $T3$, and $T4$ are a distance h from $T0$.

The law of conservation of energy in this case says that the net heat flowing into ABCD must equal the net heat flowing out, as energy cannot be created out of nothing. The heat flowing across a boundary of ABCD is proportional to the temperature gradient there. Thus conservation of energy says

$$\left(\begin{array}{c}\text{Heat in}\\\text{across AC}\end{array}\right)+\left(\begin{array}{c}\text{Heat in}\\\text{across CD}\end{array}\right)=\left(\begin{array}{c}\text{Heat out}\\\text{across AB}\end{array}\right)+\left(\begin{array}{c}\text{Heat out}\\\text{across BD}\end{array}\right).$$

The heat coming in across AC is (approximately) proportional to temperature gradient $(T0 - T1)/h$, and also to the length h of AC, and so on. Thus:

$$\frac{T0-T1}{h}\times h+\frac{T0-T4}{h}\times h=\frac{T2-T0}{h}\times h+\frac{T3-T0}{h}\times h\,,$$

or rearranging $T0 = \frac{1}{4}(T1 + T2 + T3 + T4)$.

The law of conservation of energy gives the mathematical rule that temperature at one node = average of temperatures at the four nearest surrounding nodes.

DO THIS NOW! 13.9

Think about and explain why this is approximate. Do not read on until you have thought about this. **STOP AND THINK NOW.**

It is approximate because:

- *our expression for the temperature gradient is only approximate – remember how the derivative is formed – h must be small and in the limit tend to zero to get an exact value for the gradient.*

- *saying that the heat flow is proportional to the length of side assumes that the gradient is constant all along that side, which of course it isn't.*

What is evident is that the smaller h is (that is, the finer the overlaid grid is) the more accurate our answer will be.

13.6 1 Finite differences in action

To illustrate the method in action we shall solve the problem shown in Figure 13.4, using the coarsest possible grid as shown there.

Example 13.6.1

The flat plate as shown in Figure 13.4 has the temperature on three of its edges held at $T = 0$, with the fourth edge held at a temperature of $T = 90°$ (Figure 13.6).

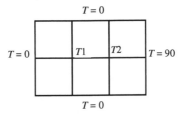

Figure 13.6 Setting up the temperature problem with the coarsest grid possible.

An approximate value of the temperature at certain points is obtained as follows:

- Place a regular *square* grid over the plate as shown.

- Name the temperatures at the two inside nodes of the grid $T1$ and $T2$.

- Then according to the finite difference model above, the values of $T1$ and $T2$, which are approximate values of the temperature at those points, can be found by noting that the temperature at each node is the average of the temperatures at the four surrounding nodes.

This results in two simultaneous equations, which can be solved for $T1$ and $T2$. The equations are:

$$T1 = \frac{1}{4}(0 + 0 + 0 + T2) \quad \text{and} \quad T2 = \frac{1}{4}(0 + T1 + 0 + 90) .$$

Rewriting these equations in standard form:

$$4T1 - T2 = 0$$
$$-T1 + 4T2 = 90$$

Solving these simple equations (DO THIS NOW!):

$$T1 = 6$$
$$T2 = 24$$

Apart from checking your working you can also ask if these answers are sensible and obey engineering common sense. For instance, do they lie between 0 and 90, and do they get hotter towards the right hand end of the plate? The answer is yes. As an engineer you should always be looking to see if your answers are reasonable – for instance if you made an error with the arithmetic here and got a temperature of 10000 then we would expect you to notice that all is not well!

Example 13.6.2 Refining the grid

To get a more accurate figure for the temperature distribution across the plate, you could put a finer grid over the plate. The next finest square grid is that shown in Figure 13.7. Temperatures at the nodes of this grid can then be estimated by noting that each temperature is again the average of the four surrounding points. You might notice that two of these points, marked by circles, correspond to the points of the first, coarser grid, so we will have a basis for comparison.

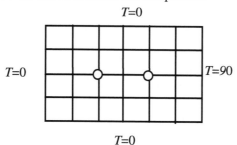

Figure 13.7 Covering the region with a finer grid.

DO THIS NOW! 13.10

Think briefly about each of these questions *before* reading the response to each.

At how many points must the temperatures now be calculated?

There are now 15 internal points so we need temperatures at 15 points, which means we must solve 15 simultaneous equations.

Can symmetry help?

The answer is that there is symmetry across the central horizontal line in this problem, so that the temperatures on the top line must equal those on the bottom line. There are thus only 10 temperature values to find.

What are the equations for the temperatures on this refined grid? Once more you get these by using the averaging rule. First label the temperatures, as partially shown in Figure 13.8.

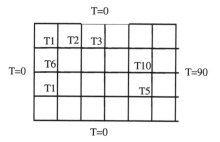

Figure 13.8 The finer temperature grid in more detail.

Thus the equation for $T1$ is $\qquad\qquad\qquad T1 = \frac{1}{4}(0+0+T6+T2)$.

DO THIS NOW! 13.11

The 9 other equations can be similarly generated for the other 9 points. Find these other 9 equations now. Leave them for now in the form $T1 = ..., T2 = ...,$ etc.

Now how would you solve these equations? You might answer that you would use your technology, but you can see that the number of equations is beginning to grow. For instance if you try to use your graphic calculator, you have a lot of algebraic rewriting to get the equations into the right shape. You will also run off the screen and it becomes difficult to keep track.

In terms of method, it is most natural to use the Gauss-Seidel iterative method, or something like it, to solve the equations, especially since as they are formulated, the first equation automatically gives $T1$ in terms of the other variables, the second equation gives $T2$ in terms of the other variables and so on.

We now move to an open-ended approach, and suggesting various things for you to do, but without detailed instructions. Enjoy seeing how well you can handle this.

DO THIS NOW! 13.12

Look into using the Gauss-Seidel method to solve the 10 equations you have generated from Example 13.6.2. If you attempt this on paper, you will quickly find the calculations very burdensome, repetitive and open to error.

We suggest you think about ways in which you can make a spreadsheet such as Excel do the repetitive Gauss-Seidel calculations for you, using just spreadsheet formulae to implement the algorithm.

Check any answers you obtain by whatever means you can: comparison with the finer grid case, engineering common sense, sample checks back into the original equations, and so on.

Look into the possibilities offered by any other technology available to you; CAS systems often have simultaneous equation solvers, but can be open to the same issues as graphic calculators; often engineers will write their own software to implement numerical methods, sometimes calling on standard commercial software libraries such as NAG (Numerical Algorithms Group).

DO THIS NOW! 13.13

Sometimes there are special possibilities in certain technologies. For instance, Figure 13.9 shows a spreadsheet set up to use Excel's "iteration" feature to implement the Gauss-Seidel method as required in DO THIS NOW 13.12! Study Figure 13.9 and see if you can set up a spreadsheet like this for yourself. Here are some hints:

The spreadsheet mirrors the *physical* shape of the area, with each node temperature appearing in a cell in its corresponding position.

Each cell formula makes its value equal to the average of the four surrounding cells. This will lead to "circular references", but they can be removed as follows.

You must implement iteration by Gauss-Seidel, by choosing "Calculation" from the "Tools" menu and changing mode to "Iteration" as shown.

You can plot temperature profile across the plate as shown using a "Surface plot".

With this technological support, you can easily refine the grid further by using more cells. The number of equations increases rapidly but the spreadsheet deals with that! Note the current limit on number of columns in Excel is 256 (!).

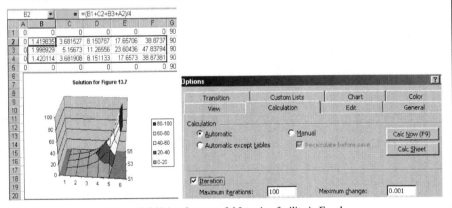

Figure 13.9 Using the powerful Iteration facility in Excel.

END OF CHAPTER 13 EXERCISE – DO THIS NOW!

For the equation systems (a) and (b) below, solve them by:

(i) Gaussian elimination (with back substitution);

(ii) Gauss-Jordan elimination;

(iii)   Gauss-Seidel iteration;

(iv)   using more than one technology – e.g. graphic calculator, CAS, spreadsheet.

(a)
$$7x_1 - x_2 + 2x_3 = 7$$
$$-2x_1 + 9x_2 + x_3 = 3$$
$$x_1 - 2x_2 + 8x_3 = 10$$

(b)
$$10x_1 + x_2 + 2x_3 - x_4 = 3$$
$$-2x_1 + 11x_2 + x_3 + 3x_4 = 5$$
$$2x_1 - x_2 + 9x_3 + x_4 = 10$$
$$-2x_1 + 2x_2 + 8x_3 - 10x_4 = 1$$

Check that your answers agree in each case. Compare the pros and cons of the different approaches in each case.

14

Vectors

"It's all visual. It's hard to explain." Richard Feynman

14.1 INTRODUCTION

The notion of a *vector* allows you to deal with physical quantities which have not only a magnitude but a direction as well. If a quantity has magnitude only it is called a *scalar*. If it has both magnitude and direction it is called a *vector*.

Words used imprecisely in everyday language need to be used precisely once this distinction is recognized. For example, velocity is a vector, being speed in a given direction; speed is a scalar, measuring how fast but not specifying direction. Weight is the action of mass in a given direction (under gravity) and is thus a vector; mass is the tendency of a body to resist force, and is a scalar. Other examples of scalars are time and temperature; other examples of vectors are position, acceleration, and force.

A scalar can be specified by a single numerical value together with its units; vectors are specified by a numerical size and a direction in space. In a two-dimensional space such as a plane, two numbers are needed to specify direction; in three dimensions three numbers are needed to specify direction.

Vector theory is important in engineering. It allows vector quantities to be handled mathematically in very concise ways so that complexity in the mathematics does not obscure the engineering. You must learn how to handle these mathematical objects. This chapter gives you a simple introduction. **Do the exercises and activities** to become familiar with vectors.

14.2 REPRESENTING VECTORS

14.2.1 Basic Cartesian notation

You can represent a vector geometrically by an arrow pointing in the right direction, with its length proportional to the size or magnitude of the vector. The vector can also be represented algebraically, in the same way as a point in space, by the co-ordinates of the end of the arrow relative to the origin O. In two dimensions we need x and y co-ordinates, called the *components* of the vector; in three dimensions we need x, y and z components. You can see the two-dimensional case in Figure 14.1, where the position of a point P, which is a vector, is represented as (x, y). We denote a vector by a bold, underlined lower case letter. Thus the position vector of the point P is labelled as $\underline{r} = (x, y)$.

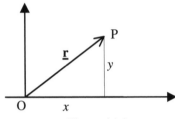

Figure 14.1

Because the vector starts at the origin O and ends at the point P, another notation often used is shown in (14.1), where the arrow shows the direction of the vector.

$$\underline{r} \equiv \overrightarrow{OP} \qquad (14.1)$$

14.2.2 Magnitude and unit vectors

The size of the vector $\underline{r}$, as represented by the length of the arrow, is called the *magnitude* of $\underline{r}$ and is written $|\underline{r}|$.

It is useful to have the idea of a *unit vector*, written $\hat{\underline{r}}$, which is the vector of magnitude 1, pointing in the same direction as $\underline{r}$. The unit vector $\hat{\underline{r}}$ is then related to $\underline{r}$ by

$$\hat{\underline{r}} = \frac{\underline{r}}{|\underline{r}|} \qquad (14.2)$$

You can see from Figure 14.1 that the magnitude of a two-dimensional vector is the length of the line OP, which is given in terms of x and y by Pythagoras' theorem

$$|\underline{r}| = \sqrt{x^2 + y^2} \qquad (14.3)$$

In 3 dimensions, magnitude is similarly given by

$$|\underline{r}| = \sqrt{x^2 + y^2 + z^2} \qquad (14.4)$$

You will notice that there are many parallels between two-dimensional vectors and the ideas of complex numbers, which you met in Chapter 6.

14.2.3 Direction cosines

Another very useful associated idea involves the direction cosines of a vector. These are useful in many ways, for example in robotics, and in finding the lines of principal stresses and strains in stress analysis.

The *direction cosines* are the cosines of the angles made by the vector with the three Cartesian axis directions x, y and z. For the three-dimensional vector $\underline{r} \equiv (x, y, z) \equiv x\hat{\underline{i}} + y\hat{\underline{j}} + z\hat{\underline{k}}$ they are given respectively by:

$$l = \cos(\alpha) = \frac{x}{|\underline{r}|} \qquad m = \cos(\beta) = \frac{y}{|\underline{r}|} \qquad n = \cos(\gamma) = \frac{z}{|\underline{r}|}$$

The letters l, m and n are conventionally used to represent direction cosines, and the idea is illustrated geometrically in Figure 14.2.

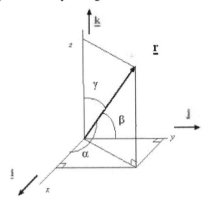

Figure 14.2

14.3 THE ALGEBRA OF VECTORS
14.3.1 Vector addition

Two vectors can be added both geometrically and algebraically, to form what is known as their resultant. The geometrical approach relies on the use of a *vector triangle* as shown in Figure 14.3. The two vectors to be added are placed head to tail. This forms two sides of a triangle. The third side of the triangle forms the *sum* or *resultant*.

In Figure 14.3 for instance, $\underline{r}_3 = \underline{r}_1 + \underline{r}_2$ or $\overrightarrow{AC} = \overrightarrow{AB} + \overrightarrow{BC}$.

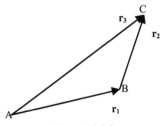

Figure 14.3

Algebraically, to add two vectors, you add each component individually. Thus if $\underline{a} = (a_1, a_2, a_3)$ and $\underline{b} = (b_1, b_2, b_3)$, then $\underline{a} + \underline{b} = (a_1 + b_1, a_2 + b_2, a_3 + b_3)$.

14.3.2 Multiplication by a scalar

If a vector is multiplied by a scalar, the rule is that every component must be multiplied by that scalar.

Thus: $k\underline{r} = k(r_1, r_2, r_3) = (kr_1, kr_2, kr_3)$

The geometrical meaning is that the length of the vector is extended by a factor k.

$k\underline{r}$ is a vector in the same direction as $\underline{r}$ but of magnitude k times $\underline{r}$.

14.3.3 Vector subtraction

One case worth mentioning is the vector $(-1)\underline{r} = -\underline{r}$, which is a vector equal in magnitude to $\underline{r}$ but in the opposite direction. This then gives us the mechanism for subtracting vectors. If you want to subtract $\underline{r}$, you simply add $-\underline{r}$.

This is shown geometrically in Figure 14.4 where, to form $\underline{r}_4 = \underline{r}_1 - \underline{r}_2$, $\underline{r}_2$ is reversed and added.

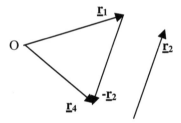

Figure 14.4

Example 14.3.1

Given the vectors $\underline{a} = (1, 2)$, $\underline{b} = (0, -1)$ and $\underline{c} = (2, 3)$, find

(a) $\underline{a} + 2\underline{b}$ (b) $\underline{a} - \underline{b}$ (c) $\underline{a} + \underline{b} - \underline{c}$ (d) a unit vector parallel to $\underline{c}$

Here are the solutions:

(a) $\underline{a} + 2\underline{b} = (1, 2) + 2(0, -1) = (1, 2) + (0, -2) = (1,0)$

(b) $\underline{a} - \underline{b} = (1, 2) - (0, -1) = (1, 3)$

(c) $\underline{a} + \underline{b} - \underline{c} = (1, 2) + (0, -1) - (2, 3) = (-1, -2)$

(d)   A unit vector parallel to $\underline{c} = \dfrac{\underline{c}}{|\underline{c}|} = \dfrac{(2,3)}{\sqrt{2^2 + 3^2}} \approx (0.15, 0.23)$

Figure 14.6 displays these results produced on CAS technology (TI Voyage 200).

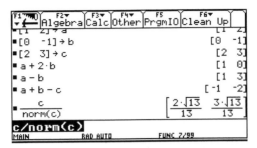

Figure 14.6

14.3.4 The standard basis vectors

Unit vectors pointing in the directions of the x, y and z axes are denoted by $\hat{\underline{i}}$, $\hat{\underline{j}}$ and $\hat{\underline{k}}$ respectively. All vectors can be represented very simply in terms of these.

In two dimensions: $\underline{r} \equiv (x, y) \equiv x\hat{\underline{i}} + y\hat{\underline{j}}$ (14.5)

In three dimensions $\underline{r} \equiv (x, y, z) \equiv x\hat{\underline{i}} + y\hat{\underline{j}} + z\hat{\underline{k}}$ (14.6)

DO THIS NOW! 14.1

Find out *now* how your technology – graphic calculator, CAS, etc – handles vectors.

1. Given vectors $\underline{a} = (4, 3)$, $\underline{b} = (-1, -2)$ and $\underline{c} = (0, 4)$, find

(a) $\underline{a} + 2\underline{b}$ (b) $3\underline{a} - 2\underline{b}$ (c) $\underline{a} - \underline{b} - \underline{c}$ (d) a unit vector parallel to $\underline{b}$

2. Given the vectors $\underline{a} = (1, 1, 1)$, $\underline{b} = (-1, 2, 3)$ and $\underline{c} = (0, 3, 4)$, find

(a) $\underline{a} + \underline{b}$ (b) $2\underline{a} - \underline{b}$ (c) $\underline{a} + \underline{b} - \underline{c}$ (d) a unit vector parallel to $\underline{c}$

Check your own answers using your technology.

14.4 PRODUCTS OF VECTORS

There are two kinds of products of vectors: the *scalar* or *dot* product and the *vector* or *cross* product. They are each defined the way they are because they have particular uses when vectors are used in engineering and science.

14.4.1 The scalar or dot product

The *scalar* or *dot product* of two vectors $\underline{a} = (a_1, a_2)$ and $\underline{b} = (b_1, b_2)$ is defined in two equivalent ways, one algebraic and one geometric, as follows:

Algebraically: $\underline{a} \cdot \underline{b} = a_1 b_1 + a_2 b_2$

$(\underline{a} \cdot \underline{b} = a_1 b_1 + a_2 b_2 + a_3 b_3$ in 3D)

Geometrically: $\underline{a} \cdot \underline{b} = |\underline{a}||\underline{b}| \cos\theta$

$(\theta,\ (0 \le \theta \le \pi)$, is the angle between the two vectors.)

$\underline{a} \cdot \underline{b}$ is a *scalar* (just a number). The dot between the vectors is an essential part of the notation. The definitions *are* equivalent, although we shall not prove that here.

There are some simple but useful rules obeyed by the dot product:

$$\underline{u} \cdot \underline{v} = \underline{v} \cdot \underline{u} \qquad (14.7)$$

$$\underline{u} \cdot \underline{u} = |\underline{u}|^2 \text{ since in this case } \theta = 0. \qquad (14.8)$$

$$\underline{u} \cdot (\underline{v} + \underline{w}) = \underline{u} \cdot \underline{v} + \underline{u} \cdot \underline{w} \qquad (14.9)$$

Example 14.4.1

Given the vectors $\underline{a} = (0, -2)$, $\underline{b} = (3, 4)$ and $\underline{c} = (2, 1)$, evaluate

(a) $\underline{a} \cdot \underline{c}$ (b) $\underline{b} \cdot \underline{c}$ (c) $(\underline{a} + \underline{b}) \cdot \underline{c}$ (d) $\underline{a} \cdot (2\underline{b} + 3\underline{c})$ (e) $(\underline{a} \cdot \underline{b})\underline{c}$

Solutions

(a) $\underline{a} \cdot \underline{c}$ $= (0 \times 3 + -2 \times 1) = -2$

(b) $\underline{b} \cdot \underline{c}$ $= (3 \times 2 + 4 \times 1) = 10$

(c) $(\underline{a} + \underline{b}) \cdot \underline{c}$ $= (0+3,-2+4) \cdot (2,1) = (3,2) \cdot (2,1) = 8$

(d) $\underline{a} \cdot (2\underline{b} + 3\underline{c})$ $= (0,-2).(6+6,8+3) = (0,-2).(12,11) = -22$

(e) $(\underline{a} \cdot \underline{b})\underline{c}$ $= (0 \times 3 + (-2) \times 4) \underline{c} = (-16,-8)$ (Note this is a *vector!*)

Figure 14.8 shows some of Example 14.4.1 carried out on a CAS enabled graphic calculator.

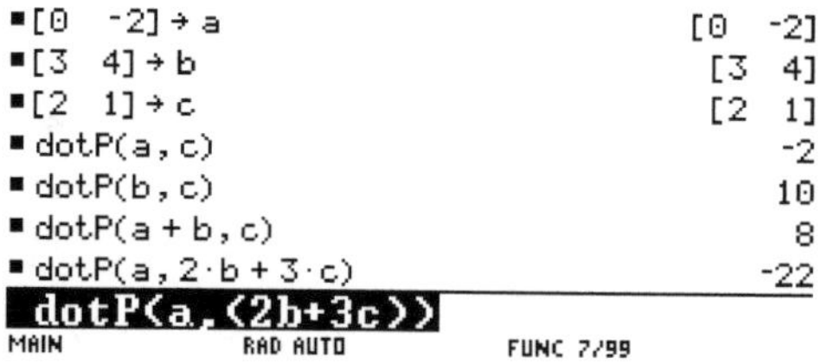

```
■[0  -2] → a                        [0  -2]
■[3   4] → b                        [3   4]
■[2   1] → c                        [2   1]
■ dotP(a, c)                              -2
■ dotP(b, c)                              10
■ dotP(a + b, c)                           8
■ dotP(a, 2·b + 3·c)                     -22
 dotP(a, (2b+3c))
MAIN            RAD AUTO        FUNC 7/99
```

Figure 14.8

DO THIS NOW! 14.2

1. Verify rules (14.7) – (14.9) for $\underline{u} = (1, 2, 3)$, $\underline{v} = (-1, 4, 2)$, $\underline{w} = (-2, -3, 5)$.

2. Given the vectors $\underline{a} = (-1, 2)$, $\underline{b} = (0, 6)$ and $\underline{c} = (5, -5)$, evaluate

(a) $\underline{a} \cdot \underline{c}$ (b) $\underline{b} \cdot \underline{c}$ (c) $(\underline{a} - \underline{b}) \cdot \underline{c}$ (d) $\underline{a} \cdot (4\underline{b} + \underline{c})$ (e) $(\underline{a} \cdot \underline{b})\underline{c}$

14.4.2 Uses of the dot product

Some of the properties of the dot product have special uses, as in these examples.

Perpendicular vectors

The geometrical form of the dot product includes a term $\cos(\theta)$. If $\theta = 90°$ then $\cos(\theta) = 0$, and so the dot product will be zero. Therefore, providing $\underline{u}$ and $\underline{v} \neq 0$, then $\underline{u} \cdot \underline{v} = 0$ tells you that $\underline{u}$ and $\underline{v}$ are perpendicular.

Resolving vectors into x and y components

It is useful to be able to find the component of the vector **a** along a given direction – for instance to see how much of a force pushes in the *x* or *y* directions. From Figure 14.7, you can see from the right-angled triangle that the component of vector **a** in the direction of OP is given by $ON = |a|\cos\theta$.

If we take a *unit* vector along OP to be $\hat{\mathbf{n}}$ then $|\hat{\mathbf{n}}| = 1$ and ON can be written:

$$ON = |a|\cos\theta = |a||\hat{n}|\cos\theta = \mathbf{a}\cdot\hat{\mathbf{n}}$$

Thus the component of vector **a** in the direction defined by the unit vector $\hat{\mathbf{n}}$ is given by $\mathbf{a}\cdot\hat{\mathbf{n}}$.

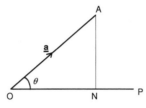

Figure 14.7

Example 14.4.2

Given the vectors $\underline{a} = (1, 2)$, $\underline{b} = (-1, 1)$ and $\underline{c} = (2, 2)$:

   (a) Evaluate the angle between $\underline{a}$ and $\underline{b}$

   (b) Identify if any of $\underline{a}$, $\underline{b}$ and $\underline{c}$ are perpendicular to each other.

(a)$\underline{a}\cdot\underline{b} = |a||b|\cos\theta = a_1b_1 + a_2b_2$

      Now $|a| = \sqrt{5}$ and $|b| = \sqrt{2}$ and $\underline{a}\cdot\underline{b} = 1\times(-1) + 2\times1 = 1$

      Thus $1 = \sqrt{5}\times\sqrt{2}\times\cos(\theta)$ and so

$$\theta = \cos^{-1}(\frac{\mathbf{a.b}}{|a||b|}) = \cos^{-1}(\frac{1}{\sqrt{10}}) = 1.249 \text{ radians to 3 d.p.}$$

(b) $\underline{b}$ and $\underline{c}$ are perpendicular to each other.

      You can check both answers by drawing the picture.

DO THIS NOW! 14.3

Find out now how your technology implements the dot product, and test yourself out by reproducing the results of Example 14.4.1 and **DO THIS NOW 14.2!**

Use the dot product to find the component of the vector $\mathbf{F} = (1, -2)$ in

(a)   the $\hat{\mathbf{i}}$ direction   (b)   the direction (3/5, 4/5)

(c) the direction (2, -1)

14.4.3 The vector or cross product

The other type of product between two vectors **a** and **b** is the **vector** or **cross product.** Its result is a *vector* which is at right angles to both **a** and **b**. It thus necessarily involves the third dimension. One use is in describing moments of forces in mechanical engineering problems.

The product is again defined in two equivalent ways, one geometric, and one algebraic in terms of components. It is best to meet the geometrical definition first.

Geometric definition of cross product

For two vectors **a** and **b** the *vector* or *cross product* is defined geometrically as

$$\underline{a} \times \underline{b} = |\underline{a}||\underline{b}| \sin \theta \, \hat{n}$$

where θ is the angle between **a** and **b** ($0 \le \theta \le \pi$), and $\hat{n}$ is the unit vector perpendicular to both **a** and **b** pointing in the direction such that **a**, **b** and $\hat{n}$ form a right-handed set, i.e. if θ is measured from **a** to **b**, the direction of the vector product **a**×**b** is the direction in which a right-handed screw would advance when turned from **a** to **b**.

As with the dot product there are some simple rules.

$\underline{a} \times \underline{b} = -\underline{b} \times \underline{a}$ (reversing **a** to **b** reverses direction of $\hat{n}$)   (14.10)

$\underline{a} \times \underline{a} = 0$ (because the angle θ between **a** and itself is 0)   (14.11)

$\underline{a} \cdot (\underline{a} \times \underline{b}) = 0$ (because **a** is perpendicular to **a**×**b**)   (14.12)

$\underline{a} \times (\underline{b} \times \underline{c}) \ne (\underline{a} \times \underline{b}) \times \underline{c}$ in general   (14.13)

$\underline{a} \times (\underline{b} + \underline{c}) = (\underline{a} \times \underline{b}) + (\underline{a} \times \underline{c})$ (14.14)

Algebraic definition of cross product

The equivalent algebraic definition of cross product in terms of vector components is obtained by looking at the effect of the cross product on the three basic unit vectors $\hat{i}, \hat{j}$ and $\hat{k}$ defined in section 14.3.4. These satisfy:

$$\hat{i} \times \hat{i} = 0 \qquad\qquad \hat{j} \times \hat{j} = 0 \qquad\qquad \hat{k} \times \hat{k} = 0$$

$$\hat{i} \times \hat{j} = \underline{\hat{k}} \qquad\qquad \hat{j} \times \underline{\hat{k}} = \hat{i} \qquad\qquad \underline{\hat{k}} \times \hat{i} = \hat{j}$$

There are some nice patterns here. This happens often with vectors.

DO THIS NOW! 14.4

Satisfy yourself that these are true by thinking about the geometrical definition of cross product.

We can use these results to find the algebraic component form of the vector product. The derivation is complicated; you should at least aim to note the result.

Take $\underline{a} = (a_1, a_2, a_3) = a_1\hat{\underline{i}} + a_2\hat{\underline{j}} + a_3\hat{\underline{k}}$ and $\underline{b} = (b_1, b_2, b_3) = b_1\hat{\underline{i}} + b_2\hat{\underline{j}} + b_3\hat{\underline{k}}$

Then using the rules above:

$$
\begin{aligned}
\underline{a} \times \underline{b} \quad &= (a_1\hat{\underline{i}} + a_2\hat{\underline{j}} + a_3\hat{\underline{k}}) \times (b_1\hat{\underline{i}} + b_2\hat{\underline{j}} + b_3\hat{\underline{k}}) \\
&= a_1 b_1 (\hat{\underline{i}} \times \hat{\underline{i}}) + a_1 b_2 (\hat{\underline{i}} \times \hat{\underline{j}}) + a_1 b_3 (\hat{\underline{i}} \times \hat{\underline{k}}) + \\
&\quad a_2 b_1 (\hat{\underline{j}} \times \hat{\underline{i}}) + a_2 b_2 (\hat{\underline{j}} \times \hat{\underline{j}}) + a_2 b_3 (\hat{\underline{j}} \times \hat{\underline{k}}) + \\
&\quad a_3 b_1 (\hat{\underline{k}} \times \hat{\underline{i}}) + a_3 b_2 (\hat{\underline{k}} \times \hat{\underline{j}}) + a_3 b_3 (\hat{\underline{k}} \times \hat{\underline{k}}) \\
&= a_1 b_2 \hat{\underline{k}} + a_1 b_3 (-\hat{\underline{j}}) + a_2 b_1 (-\hat{\underline{k}}) + a_2 b_3 \hat{\underline{i}} + a_3 b_1 \hat{\underline{j}} + a_3 b_2 (-\hat{\underline{i}})
\end{aligned}
$$

Finally then

$$\underline{a} \times \underline{b} \qquad = (a_2 b_3 - a_3 b_2)\hat{\underline{i}} + (a_3 b_1 - a_1 b_3)\hat{\underline{j}} + (a_1 b_2 - a_2 b_1)\hat{\underline{k}} \qquad (14.15)$$

This looks unmemorable and complicated, but there is a way of remembering it using an extension of the idea of a determinant as you met in Chapter 12, to include 3×3 matrices. A 3×3 determinant as defined below is written in terms of three 2×2 determinants. In fact this definition works for any 3×3 determinants, but you can see the benefit in this particular case in terms of pattern: the first row is just $\hat{\underline{i}}, \hat{\underline{j}}$ and $\hat{\underline{k}}$, the second row is the components of $\underline{a}$, and the third is the components of $\underline{b}$.

$$
\underline{a} \times \underline{b} = \begin{vmatrix} \hat{\underline{i}} & \hat{\underline{j}} & \underline{k} \\ a_1 & a_2 & a_3 \\ b_1 & b_2 & b_3 \end{vmatrix} = \hat{\underline{i}}\begin{vmatrix} a_2 & a_3 \\ b_2 & b_3 \end{vmatrix} - \hat{\underline{j}}\begin{vmatrix} a_1 & a_3 \\ b_1 & b_3 \end{vmatrix} + \underline{k}\begin{vmatrix} a_1 & a_2 \\ b_1 & b_2 \end{vmatrix}
$$

DO THIS NOW! 14.5
Multiply out the 2×2 determinants in this expression and show they do give the expression (14.15) derived for $\underline{a} \times \underline{b}$.

Example 14.4.3

If $\underline{a} = (2, 2, -1)$ and $\underline{b} = (3, -6, 2)$ find $\underline{a} \times \underline{b}$.

Solution

$$
\underline{a} \times \underline{b} = \begin{vmatrix} \hat{\underline{i}} & \hat{\underline{j}} & \underline{k} \\ 2 & 2 & -1 \\ 3 & -6 & 2 \end{vmatrix} = \hat{\underline{i}}\begin{vmatrix} 2 & -1 \\ -6 & 2 \end{vmatrix} - \hat{\underline{j}}\begin{vmatrix} 2 & -1 \\ 3 & 2 \end{vmatrix} + \underline{k}\begin{vmatrix} 2 & 2 \\ 3 & -6 \end{vmatrix}
$$

$$= \hat{\underline{i}}(4 - 6) - \hat{\underline{j}}(4 + 3) + \underline{k}(-12 - 6) = (-2, -7, -18)$$

Figure 14.9 shows a check of this answer using a CAS.

Figure 14.9

DO THIS NOW! 14.6

Find out *now* how your technology handles the vector product, and test your understanding by verifying the result of Example 14.4.3.

Verify rules (14.10) to (14.14), using the algebraic definition, for $\underline{a} = (1, 2, 3)$, $\underline{b} = (-1, 4, 2)$, and $\underline{c} = (-2, -3, 5)$.

END OF CHAPTER 14 EXERCISES – DO ALL OF THESE NOW!

In each case perform calculations on paper and check by technology.

1. Given $\underline{a} = (1, 1, 0)$, $\underline{b} = (2, 2, 1)$ and $\underline{c} = (0, 1, 1)$ evaluate

    (a) $\underline{a} + \underline{b}$ (b) $\underline{a} + \underline{b}/2 + 3\underline{c}$ (c) $\underline{a} - 2\underline{b}$ (d) $|\underline{a}|$

    (e) $|\underline{b}|$ (f) $|\underline{a} - \underline{b}|$ (g) $\hat{\underline{a}}$ (h) $\hat{\underline{b}}$

    (i) $\underline{a} \cdot \underline{c}$ (j) $(\underline{a} \times \underline{b}) \cdot \underline{c}$ (k) $(\underline{a} \cdot \underline{b})\underline{c}$

2. Find the angle between the vectors $\underline{a} = (1, -3, 5)$ and $\underline{b} = (1, 2, 4)$.

3. Find the component of the vector $\underline{f} = (-2, 1, 5)$ in
    (a) the $\hat{\underline{i}}$ direction

    (b) the direction $(1/3, 2/5, 2/5)$

    (c) the direction $(0, 0, -1)$.

6. You are given $\underline{a} = (5, 4, 2)$, $\underline{b} = (4, -5, 3)$ and $\underline{c} = (2, -1, -2)$.

    (a) Find $\underline{a} \cdot \underline{b}$.

    (b) Find the angle between $\underline{a}$ and $\underline{b}$.

    (c) Find a unit vector perpendicular to $\underline{a}$ and $\underline{b}$.

    (d) Find the components of all three vectors in the direction $(2, -3, 0)$.

7. (Harder) Given the triangle OAB, where O is the origin, and denoting the mid-points of the opposite sides as O', A' and B', use vectors to show that the lines OO', AA' and BB' meet at a point. (Note that this is the result that the medians of a triangle meet at the centroid).

8. (Harder) A plane is flying at 400 knots in a strong wind, blowing from the NW at a speed of 50 knots. The plane wishes to fly due west. In what direction should it head to achieve this, and what will be the actual ground speed of the plane?

15

First Order Ordinary Differential Equations

"Trying is the first step towards failure" H Simpson

15.1 INTRODUCTION

In Chapter 10, as you explored the idea of integration further, you met the beginnings of the idea of ordinary differential equations (ODEs). These are equations involving not only the quantity you want to find, but information about its rate of change too. Such equations are important since many physical laws involve not the quantity you seek but the rate at which it changes. For instance, Newton's second law of motion states that force $F(t)$ is equal to mass m times acceleration, which is rate of change of velocity $v(t)$. The equation of motion is then $F = m\dfrac{dv}{dt}$. If you wish to know how the speed varies for a known force F, then you must solve this differential equation for the speed $v(t)$. The equation is *differential* since it involves a derivative of the unknown function $v(t)$.

Other examples of engineering situations modelled by such equations include simple cooling models, a capacitor discharging or charging through a resistor, and radioactive decay. You will explore these models further in Section 15.4.

These equations are called *ordinary* differential equations because the unknown function is a function of one variable only – we shall use time t because engineers and scientists often have to deal with systems which vary in time, although many text books insist on using x as the independent variable. This is as opposed to *partial* differential equations where the unknown function can be a function of more than variable, for instance a temperature varying in both time and position. Although these equations are not covered in this book you will have seen in Section 13.6 an example of their use and approximate solution.

In this chapter we look at simple *first order* ordinary differential equations. To call the equation first order means that only the first derivative of the unknown function occurs, and no higher derivatives. We shall mention a few of the forms these first order equations can take, and just a few of the many methods available to solve them. You will find out that solution methods vary according to the form the equation takes.

Keep on remembering to **do the exercises and activities** as you go.

15.2 OVERVIEW
15.2.1 Solving an ODE is *equivalent to* integration

The general form of a first order ordinary differential equation is $F(t, y, \dfrac{dy}{dt}) = 0$. That is, it is a mixture of terms involving the unknown function $y(t)$, its first derivative $\dfrac{dy}{dt}$ and the independent variable t. Different solution methods work in different cases depending upon the exact form of the equation.

Solving a differential equation involves finding y as a function of t, so that the behaviour of the system being modelled can be seen and understood. Whether it looks like it or not, the process of solving a first order equation is equivalent to doing one integration, since what you are doing is "getting rid of" $\dfrac{dy}{dt}$ to leave just y – and integration is the process of undoing a derivative. Indeed in some methods, for example using the Laplace transform (see Chapter 17), it seems that integration has nothing to do with what is going on – but remember that the process is equivalent.

Now because solution is equivalent to integration, an arbitrary constant will be involved, and the value of this can only be found by having extra information – typically in a time dependent system this would involve specifying the initial state of the system, say the initial temperature, position or voltage when $t = 0$. This makes sense in engineering terms – the differential equation tells you how the system varies, but to see what it is doing at time t, you must also know where it started from.

15.2.2 Types of solution method

An ***analytical solution*** to a differential equation is an algebraic relationship between y and t which satisfies the differential equation. A ***general solution*** is one which contains an arbitrary constant. A ***particular solution*** is one which satisfies a particular (perhaps initial) condition e.g. $y = 1$ when $t = 0$.

The solution methods will include: a recapitulation of the direct integration method as met in Chapter 10; the method of separation of variables, together with some engineering case studies; then a simple (so-called) approximate numerical method, implemented through a spreadsheet. Finally you will use a technological solver to get used to a very important engineering activity – producing a picture and getting a feel for how the solution behaves as the controlling parameters are varied.

15.3 DIRECT INTEGRATION REVISITED

You have already met an example using direct integration in Section 10.5, but here is a summary of the technique, in more general terms. It can be applied to any equation taking the very simple form $\dfrac{dy}{dt} = f(t)$ where $f(t)$ is any given function of time t only. The solution is given by using the definition of an integral:

$$y = \int f(t)dt + C \text{ , where } C \text{ is an arbitrary constant.}$$

Of course whether you can find an explicit algebraic solution or not, depends on whether the function $f(t)$ is one which can be integrated.

Example 15.3.1

Solve $\dfrac{dy}{dt} = \cos(t)$, subject to the initial condition $y(0) = 1$.

Integrating gives the general solution $\qquad y = \int \cos(t)dt + C = \sin(t) + C$.

But you are also given that $y(0) = 1$ (that is, $y = 1$ when $t = 0$), and so

$$1 = \sin(0) + C \text{ or } C = 1.$$

Thus the *particular* solution is $\qquad\qquad y = \sin(t) + 1$.

You can check this result on a CAS system, for instance as in Figure 15.1. Perhaps simply plot your answer and check if it satisfies your initial condition. Alternatively use a built-in solver (e.g. deSolve in Figure 15.1) and plot the result of that.

Figure 15.1 also illustrates the effect of the arbitrary constant C (represented by the @1 symbol there) by plotting some of the family of graphs produced by the general solution as C takes different values.

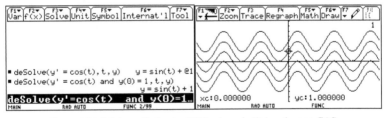

Figure 15.1 Solving a first order ODE using a built-in solver on CAS.

Example 15.3.2

Solve $\dfrac{dy}{dt} = e^{-t^2}$, subject to the initial condition $y(0) = 1$.

The theoretical solution to this equation is $y = \int e^{-t^2} dt + C$, but this integral cannot be done algebraically to give an explicit solution. In this case then we would have to take an approximate, numerical approach.

DO THIS NOW! 15.1

Solve the following first order ordinary differential equations by direct integration:

(a) $\dfrac{dy}{dt} = e^{-t}$, subject to the initial condition $y(0) = 1$.

(b) $\dfrac{dy}{dt} = -2t$, subject to the boundary condition $y(1) = 0$.

In each case plot your solution using technology and check that the picture confirms that your initial condition is satisfied.

Investigate what technological first order ODE solvers you have available to you, and check your answers here using them.

15.4 SOLUTION BY SEPARATION OF VARIABLES

Certain first order ODEs occurring commonly in engineering situations can be solved by the method of *separation of variables*. We shall illustrate this in one special and interesting case, where the rate of change of a quantity is directly proportional to its value at any point. This includes situations as diverse as cooling curves, radioactive decay, and simple discharge of a capacitor. This situation can be expressed mathematically as

$$\text{(rate of change of } y) \propto \text{ (value of } y)$$

or $$\frac{dy}{dt} \propto y \ .$$

Putting in a constant $-k$ of proportionality $\dfrac{dy}{dt} = -ky$.

(We use $-k$ since most engineering cases involve negative proportionality.)

Solving the equation means finding $y(t)$ explicitly as a function of t. This involves somehow "getting rid" of the $\dfrac{dy}{dt}$ to leave just $y(t)$, so solving is equivalent to one integration. In that case, once more there will be one arbitrary constant, and you will need another piece of information to find that constant of integration – again usually some information about how the system starts.

Here is a model solution. Read it and understand it, then absorb the conclusion.

The equation is $$\frac{dy}{dt} = -ky \ .$$

Separate the "t"s and the "y"s $$\frac{dy}{y} = -kdt \ .$$

(You might have been told previously that you should not treat the $\frac{dy}{dt}$ symbol like a fraction because it carries more meaning than that. This is true, but this is just a notational trick which works and which will help you to get to the solution.)

Integrate both sides $\qquad\qquad \int \frac{dy}{y} = \int -k\,dt$.

Do the integrals $\qquad\qquad \ln(y) = -kt + C$. (C arbitrary.)

We do not stop here, as we want y and not just $\ln(y)$. Recall that to remove a natural logarithm you use an exponential function since $\exp(\ln(mess)) = mess$!

Thus $\qquad\qquad\qquad \exp(\ln(y)) = \exp(-kt + C)$.

or $\qquad\qquad\qquad y = \exp(-kt) \times \exp(C)$.

We rewrite this using the other exponential notation (you should be comfortable with both). We also define $A = e^C$ to tidy up the involvement of the arbitrary constant. This changes nothing important – C is arbitrary and so then is A!

Thus $\qquad\qquad\qquad y = Ae^{-kt}$. $\qquad (A = e^C$ arbitrary.)

To find the arbitrary constant A you then need some more information. For example if you know that $y = y_0$ when $t = 0$, then you can write:

$$y_0 = Ae^{-k0} , \text{ which gives } A = y_0.$$

That is, A represents the initial value of y, when $t = 0$.

Conclusion: The solution of $\frac{dy}{dt} = -ky$ is an exponential curve $y = Ae^{-kt}$. The

constant A represents the initial value of y.

This does perhaps give more reason why the exponential function is so important.

Example 15.4.1

Solve $\frac{dy}{dt} = -3y$ with $y(0) = 4$.

There is no need to go through all the solution procedure every time. You should *recognize* the equation as being of the form $\frac{dy}{dt} = -ky$ and write the solution down

immediately $\qquad\qquad\qquad y = Ae^{-3t}$.

$y(0) = 4$ will give $A = 4$.

DO THIS NOW! 15.2

Solve $\quad$ (a) $\frac{dy}{dt} = -7y$ with $y(0) = 10.3$. $\qquad$ (b) $4\frac{dy}{dt} = -2y$ with $y(0) = 2$.

15.5 ENGINEERING CASE STUDIES

The type of ODE discussed in section 15.4 occurs often in modelling engineering systems in which the rate at which a quantity changes at any time is proportional to its value at that time. Important physical laws are based upon this principle.

DO THIS NOW! 15.3

There follow three simple engineering case studies, in three very different physical situations, but which all lead to the same mathematical differential equation - which is of the type discussed in section 15.4.

Work through at least one of these case studies now. Preferably work through all three, to get a feel for how the same mathematics can describe very different physical situations – a real illustration of the power of mathematics.

DO THIS NOW! 15.3.1 Case Study 1   How do things cool down?

Newton's Law of Cooling states that the rate at which an object cools is proportional to the *excess* of its temperature above the surrounding or ambient temperature.

Thus if θ is the *excess* temperature, then this is expressed mathematically as:

$$\frac{d\theta}{dt} = -k\theta \qquad \text{with } -k \text{ the constant of proportionality.}$$

Write down the general solution of this equation by separating the variables and doing the integrals which arise. Explore graphically the effect of varying the parameter k. Use any technology you wish.

Put the kettle on

A cup of tea is made with boiling water. Take the moment it is made as $t = 0$.

The average surrounding room temperature is 20°C. After 10 minutes the tea has cooled to a temperature of 80°C. Use the exponential model derived from Newton's Law of Cooling to find out (graphically or algebraically):

(i)   what the tea's temperature is after 15 minutes;

(ii)   how long it will be before it reaches a temperature of 50°C.

On a separate occasion, three cups of tea are made from the same kettle of boiling water, at almost exactly the same time. After 10 minutes, one cup has cooled to 80°C, one to 75°C, and the third to 70°C. What reasons can you think of to explain the differences?

What time did the victim die?

A murder victim is discovered by the police at 6:00a.m. The body temperature of the victim is measured then and found to be 25°C. A doctor arrives on the scene of the crime 30 minutes later and measures the body temperature again. It is found to be 22°C. The temperature of the room has remained constant at 15°C. The doctor, knowing that normal body temperature is 37°C, is able to estimate the time of death of the victim. What would be your estimate for the time of death? How accurate would you say your estimate is? (We got between 4.53 and 4.54a.m. Do you agree?)

DO THIS NOW! 15.3.2 Case Study 2 Radioactivity: How old is Pete Marsh?

When an organism is alive it maintains a mix of radioactive Carbon 14 and stable Carbon 13 which means it has a steady radioactivity level of 6.53 pico-curies per gram. When it dies this level starts to decay. Measuring how far the level has decayed provides the basis for Radiocarbon Dating. Let $R(t)$ be the radioactivity of a gram of carbon inside a dead organism (plant or animal) t years after death. The rate of decay of radioactivity R per year is

$$\frac{1}{8300} \times R .$$

Ernest Rutherford in 1908 first expressed this decay mathematically as the differential equation

$$\frac{dR}{dt} = -\frac{1}{8300} R .$$

Use Rutherford's equation together with your knowledge of the radioactivity at time $t = 0$ (i.e. time of death) to find R as a function of t.

Now try to answer this question:

Pete Marsh was the name given to some ancient human remains found well preserved in a bog in Cheshire. Carbon 14 radioactivity levels were measured at about 5.3 pico-curies per gram. Roughly how old is the body? (We get 1732 years.)

How sensitive is your estimate of the date to the accuracy of the reading of the radioactivity level? How accurate is the radiocarbon dating process, and what are the practical constraints on using it?

DO THIS NOW! 15.3.3 Case Study 3 How does a flashgun work?

Investigate how a capacitor can discharge electrical energy causing a brief sharp current to flow. Start with a much simpler circuit than that in a real flashgun, by analyzing the circuit in Figure 15.2 using Kirchoff's Laws. Show that the equation for the voltage $v(t)$ across the capacitor in terms of C, R and the applied e.m.f $E(t)$ is

$CR\dfrac{dv}{dt} + v = E(t)$. (You may need to use the capacitor law $C\dfrac{dv}{dt} = i$.)

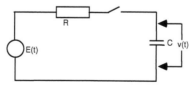

Figure 15.2 A simplified discharger circuit.

To find out how the circuit discharges the capacitor from an initial voltage of say 3 volts, try taking an applied e.m.f $E(t) = 0$, and $v(0) = 3$.

Draw some pictures of the discharging curve, and experiment with the effect of changing C and R - or the so-called *time constant CR*.

Find out what you can about real flash gun circuits, and compare what you find about them with the analysis you have performed here.

15.6 NUMERICAL SOLUTION METHODS – THE EULER METHOD
15.6.1 The need for numerical solutions

We have seen that some commonly occurring first order differential equations have analytical solutions. For example $\dfrac{dy}{dt} = -0.5y$ with initial condition $y(0) = 1$, has the solution $y = e^{-0.5t}$.

However many real situations are modelled by using differential equations of the general form $\dfrac{dy}{dt} = f(t, y)$ where $f(t, y)$ is a function of t and y, for which, unfortunately, analytical solutions may be difficult to compute, or even do not exist at all. In these cases numerical methods may be used to provide a solution. A **numerical solution** to a differential equation is a numerical approximation of a y-value for a given t-value. Although these solutions are often called approximate, and in that way seen as inferior, given modern computing power, they can often be made as accurate as required for engineering purposes.

To give an idea of how the ideas behind some of these methods work, we shall look at just the simplest numerical method – Euler's method.

15.6.2 Euler's method

Suppose we are told that $y = y_0$ when $t = t_0$, as a starter condition, where y_0 and t_0 are given constant values. The general idea is that we generate approximate values of y, which we call y_1, y_2, y_3, ..., at successive equally spaced values of t spaced a distance h apart, which we call t_1, t_2, t_3, ... This situation is shown in Figure 15.3.

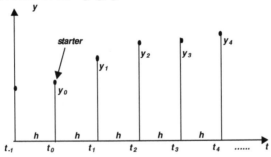

Figure 15.3 The set-up for Euler's method.

Specifically we define $t_1 = t_0 + h$, $t_2 = t_1 + h$, and in general $t_{n+1} = t_n + h$.

We know the y value y_0 corresponding to the first t value t_0. To calculate the subsequent y values corresponding to the other t values, we use an approximate form of the differential equation, $\dfrac{dy}{dt} = f(t, y)$. In fact we approximate the derivative at time t_n by

$$\frac{dy}{dt} \approx \frac{\text{difference in } y \text{ values}}{\text{difference in } t \text{ values}} = \frac{y_{n+1} - y_n}{t_{n+1} - t_n} = \frac{y_{n+1} - y_n}{h}.$$

Using this, an approximate form of the differential equation is then

$$\frac{y_{n+1} - y_n}{t_{n+1} - t_n} = f(t_n, y_n)$$

which can be rewritten $y_{n+1} = y_n + hf(t_n, y_n)$.

Thus if we know what y is at t_n we can use this formula in a systematic way to find y at the next value t_{n+1} and so on. And since we know that y is y_0 at our starting point t_0, then we can calculate y by stepping forward (or back!) in time from there.

This is the process known as Euler's method. Before following through an example it is interesting to see what is happening geometrically, as in Figure 15.4.

For the first step, the Euler equation can be written $y_1 = y_0 + hf(t_0, y_0)$, and then rewritten again as $\dfrac{y_1 - y_0}{h} = f(t_0, y_0)$. But $f(t_0, y_0)$ is just the slope of the solution curve at t_0. Thus $\dfrac{y_1 - y_0}{h} \approx \dfrac{dy}{dt}$ at t_0.

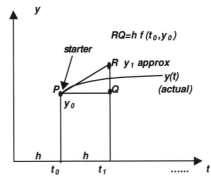

Figure 15.4 The geometrical reasoning behind Euler's method.

In terms of Figure 15.4, then, if PR is the tangent at time t_0 then QR is $y_1 - y_0$, and instead of calculating values of y by "going up" the "exact" solution curve $y(t)$, we are predicting the next value by going up the tangent to the solution curve.

15.6.3 An example of Euler's method in action

One good way to implement this method is through a spreadsheet. It can also be implemented through a simple computer or graphic calculator program, but the spreadsheet makes what is going on very explicit.

Example 15.6.1

Solve the differential equation $\dfrac{dy}{dt} = \dfrac{1}{1+t^2}$ with the initial condition $y = 0$ when $t = 0$.

We choose this particular example because an analytical solution is possible, and can be obtained by direct integration. This will give a benchmark against which you can compare the numerical solution to see how well it works, and to see what the issues are when you use the numerical solver.

DO THIS NOW! 15.4

Show by direct integration (tables or technology) that the analytical solution of $\dfrac{dy}{dt} = \dfrac{1}{1+t^2}$ with $y = 0$ when $t = 0$ is given by $y = \arctan(t)$.

Now we shall set up the Euler method. In this case $t_0 = 0$, $y_0 = 0$ and $f(t, y) = \dfrac{1}{1+t^2}$.

We shall generate values for y with a time step h of size 0.1 as shown in the spreadsheet in Figure 15.5. The formulae used are:

$$t_0 = 0, \quad y_0 = 0 \quad \text{and} \quad t_{n+1} = t_n + h \quad (n = 0, 1, 2, 3, \ldots)$$

$$y_{n+1} = y_n + hf(t_n, y_n) = y_n + h\left(\frac{1}{1+t_n^{\,2}}\right).$$

The Euler solution appears in Figure 15.5, and the analytical solution in Figure 15.6.

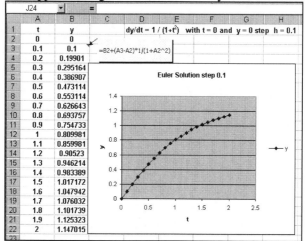

Figure 15.5 Solution of Example 15.6.1 by Euler's Method

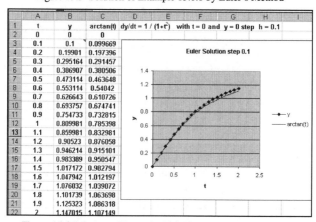

Figure 15.6 Solution of Example 15.6.1 by separation of variables

At early values of time, the numerical solution stays close to the exact solution, but as time goes on, it drifts further and further away.

DO THIS NOW! 15.5

Recreate the spreadsheets in Figures 15.5.and 15.6.

Your numerical solution will become more accurate if you use a smaller time step, h. For instance, if you halve the time step, h, to a value of 0.05, you get the situation in Figure 15.7. The approximate solution stays close to the exact solution for longer. Indeed, one strategy in using this kind of method is to use this "interval halving" technique – to go on halving the time step until the solution changes no more to within the precision required over the range of values that the solution is needed.

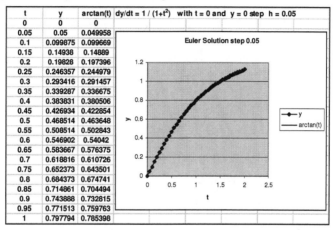

t	y	arctan(t)	dy/dt = 1 / (1+t²) with t = 0 and y = 0 step h = 0.05
0	0	0	
0.05	0.05	0.049958	
0.1	0.099875	0.099669	
0.15	0.14938	0.14889	
0.2	0.19828	0.197396	
0.25	0.246357	0.244979	
0.3	0.293416	0.291457	
0.35	0.339287	0.336675	
0.4	0.383831	0.380506	
0.45	0.426934	0.422854	
0.5	0.468514	0.463648	
0.55	0.508514	0.502843	
0.6	0.546902	0.54042	
0.65	0.583667	0.576375	
0.7	0.618816	0.610726	
0.75	0.652373	0.643501	
0.8	0.684373	0.674741	
0.85	0.714861	0.704494	
0.9	0.743888	0.732815	
0.95	0.771513	0.759763	
1	0.797794	0.785398	

Figure 15.7 Smaller time step equals more accuracy in Euler's method.

DO THIS NOW! 15.6

Show by separating the variables that the analytical solution of the first order ODE

$$\frac{dy}{dt} = y$$ (which models unlimited *growth* – rate of change of y is equal to y, and so

the bigger y is, the faster it grows!) is $y = y_0 e^t$.

Modify your spreadsheet as created in **DO THIS NOW! 15.5** to solve the same ODE approximately using Euler's method.

Start with a time step of 0.1. Comment on how well the Euler method approximates to the exact solution. Implement a strategy of time step halving, and comment on how quickly the accuracy improves.

15.7 EXPLORING THE PARAMETERS

A very useful skill for engineers is to get a feel for how the values of the parameters (or constants) in a differential equation affect the nature of the solution in that equation. With modern technological solvers, particularly if they give graphical output, this skill can be readily acquired. You may use a numerical solver on a spreadsheet, or built-in solvers in a CAS system to help you with this.

DO THIS NOW! 15.7

Find out what technological solvers - CAS or otherwise - you have available. You should at least have the Euler spreadsheet you created for **DO THIS NOW 15.5!**

Use a solver to investigate and describe in writing how the solution of the equation $\dfrac{dy}{dt} = -\alpha y$ changes as the parameter α varies (take perhaps $y = 1$ when $t = 0$):

Make your descriptions particularly in terms of the solution as presented graphically. Describe what happens to the graph in the ranges $\alpha = 0$, $\alpha > 0$, and $\alpha < 0$, and also what happens as α becomes more positive, or more negative. Do this before reading the *Hint* below this box – then compare your comments with ours.

(*Hint: you should see that if $\alpha < 0$ then the solution describes exponential decay from the initial value of y down to zero, and that as α gets more negative, the rate of the decay increases – that is the decay happens in a shorter time. If $\alpha > 0$, then the solution describes exponential growth from the initial value of y without limit, and that as α gets more positive, the rate of the growth increases – that is the growth happens in a shorter time. If $\alpha = 0$, then the solution should just remain constant at*

a value of 1. This corresponds to the case where $\dfrac{dy}{dt} = -\alpha y = 0$ and so you would

expect no change because the rate of change is 0.)

END OF CHAPTER 15 MIXED EXERCISES – DO ALL OF THESE NOW!

1. Solve the following ordinary differential equations by direct integration, or separation of variables, as appropriate:

   (i) The constant acceleration, a, of a body is equal to its rate of change of velocity, $\dfrac{dv}{dt}$. Determine an equation for v in terms of a, u and t given that the velocity is u when $t = 0$.

   (ii) The angular velocity, ω, of a flywheel of moment of inertia, I, is given by $I\dfrac{d\omega}{dt} + N = 0$, where N is a constant. Determine ω in terms of t, given that $\omega = \omega_0$ when $t = 0$.

2. Modify your Euler's method spreadsheet to investigate how well the method works for the differential equation describing exponential decay:

   $$\frac{dy}{dt} = -y \text{ subject to the condition } y = 2 \text{ when } t = 0.$$

3. Modify your Euler's method spreadsheet to investigate how well the method works for the differential equation describing limited population growth:

   $$\frac{dy}{dt} = 3 - y \text{ subject to the condition } y = 2 \text{ when } t = 0.$$

4. Modify your Euler's method spreadsheet to investigate how well the method works for the differential equation describing logistic growth:

$$\frac{dy}{dt} = y(3 - y) \text{ subject to the condition } y = 2 \text{ when } t = 0.$$

5. For questions 2, 3, and 4 check out your answers against the analytical solutions we have obtained from a CAS system (Figure 15.8). Can you reproduce these answers from your own CAS or indeed using pencil and paper?

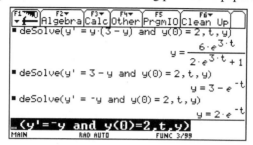

Figure 15.8 CAS output from a built-in solver.

6. Investigate and report in writing on the effects on the solutions of changing the parameter α in the differential equation $\frac{dy}{dt} = 1 - \alpha y$. Take an initial condition, say, $y = 0$ when $t = 0$.

Look particularly at what happens in the ranges $\alpha = 0$, $\alpha > 0$, and $\alpha < 0$, and describe also what happens as α becomes more positive, or more negative.

6. (*A harder modelling exercise.*) A cylindrical can has depth 10cm. and circular cross- sectional area of 6 cm^2. Water is escaping through a hole in the bottom and when the depth of water is x cm the rate of escape from it is 0.3cm^3 s^{-1}. If the can starts off full, how long will it take to empty?

16

Second Order Ordinary Differential Equations

" 'To Start Press Any Key.' Where's the ANY key?" H. Simpson

16.1 INTRODUCTION

In Chapter 15 you have just met some of the basic ideas of differential equations, through looking at first order equations. In this chapter we extend these ideas to second order equations. This is a crucial area of study for engineers, as second order, linear, constant coefficient, ordinary differential equations (O.D.E.s) provide mathematical models for engineering systems such as suspension systems, LCR electronic circuits, control systems and many other situations. Many of the ideas from first order equations carry over in a logical way.

Throughout this long chapter, remember to **do the exercises and activities**.

16.2 SECOND ORDER DIFFERENTIAL EQUATIONS

As with first order equations, we first mention a few general points before looking at how we can solve these equations. The general form of a second order linear O.D.E. with constant coefficients is

$$a\frac{d^2y}{dt^2} + b\frac{dy}{dt} + cy = f(t) \ . \tag{16.1}$$

We begin by defining the terms we use:

- *second order* means that the highest derivative of y present is second order.

- *linear* means that there is a (constant × y) term, a (constant × $\dfrac{dy}{dt}$) term and a

  (constant × $\dfrac{d^2y}{dt^2}$) term, but no more complicated terms like y^2 or $y\dfrac{dy}{dt}$ there.

- *constant coefficients* means that the quantities a, b and c are constants.

Solving the equation involves finding $y(t)$ explicitly as a function of t. As with first order equations, solution must be equivalent to an integration process, as what you are doing is getting rid of $\dfrac{dy}{dt}$ and $\dfrac{d^2y}{dt^2}$ to leave just y. The difference now is that it is equivalent to *two* integrations to get rid of a *second* derivative, and so we must expect *two* arbitrary constants to come into the answer. Thus we will require two boundary or initial conditions to find these constants.

16.3 CASE STUDY: A SUSPENSION SYSTEM

Before looking any more at the mathematics, let's have a look at one example of where such an equation comes from. Figure 16.1 shows a simple model of a suspension system, consisting of a mass connected to a fixed point by a spring and damper system, and shaken around by an external force.

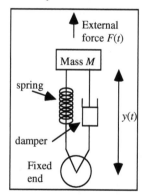

Figure 16.1 A simple representation of a suspension system.

You can analyse the vertical movement of the mass M using Newton's second law.

$$\text{Force} = \text{Mass x Acceleration} = M\frac{d^2y}{dt^2} \qquad (16.2)$$

Assume that the spring force obeys Hooke's Law:

Spring force = $-\lambda y$ (minus because if y increases, this force works to decrease it).

Assume also that the damper force is proportional to the speed of the moving mass:

Damper force = $-k\dfrac{dy}{dt}$ (again minus because the faster you move the mass, the more this opposes the motion).

Putting these into equation (16.1) leads to the equation:

$$-\lambda y - k\frac{dy}{dt} + F(t) = M\frac{d^2 y}{dt^2}$$

which can be rearranged to $\dfrac{d^2 y}{dt^2} + \dfrac{k}{M}\dfrac{dy}{dt} + \dfrac{\lambda}{M}y = \dfrac{F(t)}{M}$ (16.3)

Comparing (16.3) with the standard form in (16.1), you can see that:

"a"$=1$ Arranged by dividing through by M.

"b"$=\dfrac{k}{M}$ Thus b represents the size of the damping force.

"c"$=\dfrac{\lambda}{M}$ c represents the size of the spring force.

Note in "real" systems the parameters a, b and c are always positive.

16.4 THE SOLUTION OF LINEAR SECOND ORDER O.D.E.S
We repeat the general points we made earlier:

- All solution methods (and there are several of these) are equivalent to integrating twice, even if it does not feel as if that is what you are doing. This is because you are taking something with $\dfrac{d^2 y}{dt^2}$ in and going from there to get $y(t)$.

- Therefore the solution should always have two arbitrary constants in it, one for each integration.

- To find the value of these constants, two extra bits of information are needed - these will usually take the form of boundary or initial conditions - for example, position $y(t)$ and speed $\dfrac{dy}{dt}$ when $t = 0$.

There are then many possible approaches to solving a second order equation, each with their strengths and weaknesses; common methods are:

- *The Complementary Function/Particular Integral method.*
  This is a good way of seeing the structure of the solution, and one which helps you to understand what is happening. Technology can help with what can be heavy algebra. You will have a look at this method in Sections 16.5 to 16.8.

- *The D-operator method*, which is a bit out of fashion now.

- *The Laplace transform method.* This provides the language of control theory and of much electronics, and is one of the most widely useful transforms. It is an essential part of an engineer's knowledge, and you will meet it in Chapter 17.

- Extensions of *numerical methods* such as Euler's method (Chapter 15).

- *Direct technological solvers*, such as those built into CAS packages like Derive. These are very useful for gaining an understanding of a system's behaviour by varying the controlling parameters and seeing how the solution changes.

16.5 THE COMPLEMENTARY FUNCTION/PARTICULAR INTEGRAL APPROACH.

This is not necessarily the best *practical* approach, but it is very good for allowing you to see the structure of a solution, and therefore understanding it. The explanation is necessarily quite long, but then you can work through some examples. The method splits the solution into two parts (which have meaning both mathematically and in engineering terms) – and these can be found separately.

- The *Complementary Function* (CF) is the part of the solution which arises from unforced motion ($F(t) = 0$), that is it is the natural behaviour of the system when no external force is pushing it around. It is also called the *transient* term.

- The *Particular Integral* (PI) is an extra term added on to take account of the effect of any external force. It is also called the *steady-state* term.

Thus $y(t) = CF + PI = $ transient + steady-state

$\qquad\qquad$ = (natural system-dependent term) + (external force-dependent term).

16.6 THE CF - FINDING OUT ABOUT UNFORCED CHANGE.
16.6.1 The basis for the CF solution

We start with the CF term – which is the term reflecting the natural unforced behaviour of the system. In this case then there is no external force, and $F(t) = 0$.

Thus the CF is the solution of the equation: $a\dfrac{d^2 y}{dt^2} + b\dfrac{dy}{dt} + cy = 0$ (16.4)

How can you solve this? When faced with something like this, it is a good idea to look at a simpler case, for instance at the corresponding first order equation:

$$b\frac{dy}{dt} + cy = 0 \ . \qquad\qquad\qquad (16.5)$$

The solution of (16.5), by separation of variables, is $y(t) = Ae^{-\frac{c}{b}t}$.

DO THIS NOW! 16.1

Solve equation (16.5) by separation of variables now.

Perhaps then it is reasonable to guess that the solution to the second order case will also be exponential? Thus we shall try as a solution $y(t) = Ae^{mt}$ and substitute it into the equation (16.4) to see if it works, and if so, for what values of A or m. Here goes:

You need $y(t) = Ae^{mt}$.

Then you can find $\dfrac{dy}{dt} = mAe^{mt}$

and $\dfrac{d^2 y}{dt^2} = m^2 Ae^{mt}$.

Substituting in (16.4): $am^2 Ae^{mt} + bmAe^{mt} + cAe^{mt} = 0$.

You can divide through by Ae^{mt}, as it is not zero (since both $e^{mt} \neq 0$, and $A \neq 0$ because if it were 0 then the whole solution would disappear):

Thus $$am^2 + bm + c = 0 .$$ (16.6)

What does this say? It says that $y(t) = Ae^{mt}$ is indeed a solution to (16.4) as long as m is a root of the quadratic equation (16.6). There is no condition on A, which is therefore arbitrary. The quadratic equation (16.6) is called the *auxiliary equation*.

16.6.2 Summary of the solution for the CF

Let's see what this means in practice. Recall that the roots of a quadratic equation can be real and different, repeated real, or complex. The form of the O.D.E. solution differs depending on which of these forms the roots take. The process of finding the CF for each of these cases is long, so we present a summary of the results first so that you can see the overall simplicity, and so that the details of the derivation do not get in the way of that. The derivation follows in section 16.6.3, and examples applied to particular equations in section 16.6.4.

SOLUTION FOR THE C.F. - NO EXTERNAL FORCE

For the ODE $$a\frac{d^2y}{dt^2} + b\frac{dy}{dt} + cy = 0$$

the auxiliary equation is $am^2 + bm + c = 0$.

Case 1: $b^2 - 4ac > 0$

The damping force prevails, the roots of the quadratic equation are real and different from each other, and if their values are α and β then the solution of the o.d.e. is:

$$y(t) = Ae^{\alpha t} + Be^{\beta t}$$ system over-damped.

Case 2: $b^2 - 4ac = 0$

Transition from dominant damping, the roots of the quadratic equation are real and equal, and if the value of the root is α then the solution of the o.d.e. is:

$$y(t) = (A + Bt)e^{\alpha t}$$ system critically damped.

Case 3: $b^2 - 4ac < 0$

The damping force is relatively smaller, the roots of the quadratic equation are complex conjugates, and if the values of the roots are $p + jq$ and $p - jq$ then the solution of the o.d.e. is:

$$y(t) = e^{pt}(C\cos(qt) + D\sin(qt))$$ system under-damped.

16.6.3 The derivation of the solution for the CF

Remember the roots of the quadratic (16.6) are given by $$\text{Roots} = \frac{-b \pm \sqrt{b^2 - 4ac}}{2a}$$

and their nature is determined by the quantity $b^2 - 4ac$.

Case 1: $b^2 - 4ac > 0$

The roots will be real and different, say α and β, and so there are two possible solutions $Ae^{\alpha t}$ and $Be^{\beta t}$ (A and B are arbitrary, and not necessarily the same).

You can then generate a general solution by adding these two together (this sum certainly is a solution, and it has the required two arbitrary constants).

Thus in this case $CF = y(t) = Ae^{\alpha t} + Be^{\beta t}$.

Since $b^2 - 4ac > 0$, this means that $b^2 > 4ac$. In other words the damping force term represented by b dominates the spring force term represented by c. In this case we say the system is *overdamped*.

Case 2: $b^2 - 4ac = 0$

The roots will be real and equal, say just α, and so the only exponential solution is $Ae^{\alpha t}$. We need another arbitrary constant to make the solution complete, and it turns out that this can come from another solution, which only works for this special case, $Bte^{\alpha t}$· The general solution is formed by adding these together.

So in this case $C.F. = y(t) = Ae^{\alpha t} + Bte^{\alpha t} = (A + Bt)e^{\alpha t}$.

This is the transition from the damping force prevailing to the spring force prevailing, and we say the system is *critically damped*.

Case 3: $b^2 - 4ac < 0$

The roots will be complex (and conjugate), say $p + jq$ and $p - jq$, so in this case the solutions are $Ae^{(p+jq)t}$ and $Be^{(p-jq)t}$. The general form is the sum of these:

$$CF = y(t) = Ae^{(p+jq)t} + Be^{(p-jq)t}$$
$$= Ae^{pt}e^{jqt} + Be^{pt}e^{-jqt}$$
$$= e^{pt}(Ae^{jqt} + Be^{-jqt})$$
$$= e^{pt}(A(\cos(qt) + j\sin(qt)) + B(\cos(qt) - j\sin(qt)))$$
$$= e^{pt}((A+B)\cos(qt) + j(A-B)\sin(qt))$$
$$= e^{pt}(C\cos(qt) + D\sin(qt))$$

Note C and D are *real* and arbitrary constants. This can be confusing, but makes sense once you realize that $y(t)$ must be real as it represents a real displacement or current or voltage etc. The intermediate stages can appear complex but not the final answer. This is possible as long as A and B are complex conjugates.

In summary for Case 3: $CF = y(t) = e^{pt}(C\cos(qt) + D\sin(qt))$.

In this case the damping force is becoming less important, since the damping constant b is decreasing relative to the spring constant c, and we say the system is *underdamped*.

16.6.4 Examples of finding the CF

Although the development of the solution procedure has been long, its implementation turns out to be surprisingly simple, and only needs the summary facts as in section 16.6.2, as these examples show.

Example 16.6.1

Solve $a\dfrac{d^2y}{dt^2} + b\dfrac{dy}{dt} + cy = 0$ where $a = 1$, $b = 3$ and $c = 2$.

$b^2 - 4ac = 9 - 8 > 0$; system over-damped.

Roots are -1 and -2, both real; solution is $y = Ae^{-1t} + Be^{-2t}$.

Example 16.6.2

Solve $a\dfrac{d^2y}{dt^2} + b\dfrac{dy}{dt} + cy = 0$ where $a = 1$, $b = 2$ and $c = 1$.

$b^2 - 4ac = 4 - 4 = 0$; system critically damped.

Roots are both -1, real repeated; solution is $y = (A + Bt)e^{-1t}$.

Example 16.6.3

Solve $a\dfrac{d^2y}{dt^2} + b\dfrac{dy}{dt} + cy = 0$ where $a = 1$, $b = 2$ and $c = 2$.

$b^2 - 4ac = 4 - 8 < 0$; system under-damped.

Roots are $1 + j$ and $1 - j$, $(p = 1, q = 1)$ complex conjugate; solution is
$$y = e^{-1t}(A\cos(1t) + B\sin(1t)).$$

DO THIS NOW! 16.2

Use the summary to write down the solution of the following O.D.E.s, identifying your solutions as over-damped, under-damped or critically damped:

(Hint: check $b^2 - 4ac$; find the quadratic roots; write down the solution according to the above)

(i) $\dfrac{d^2y}{dt^2} + 5\dfrac{dy}{dt} + 4y = 0$ ($y = Ae^{-1t} + Be^{-4t}$);

(ii) $\dfrac{d^2y}{dt^2} + 4\dfrac{dy}{dt} + 4y = 0$ ($y = (A + Bt)e^{-2t}$);

(iii) $\dfrac{d^2y}{dt^2} + 4\dfrac{dy}{dt} + 5y = 0$ ($y = e^{-2t}(A\cos(1t) + B\sin(1t))$).

16.7 FINDING THE ARBITRARY CONSTANTS BY USING INITIAL CONDITIONS

Look at the solutions you have obtained in DO THIS NOW 16.2! If you wanted to plot these to get a feel for how each system behaves, you would need values for the arbitrary constants. You could just put any old value in to get a plot, but normally you will need to find these values using extra information, which frequently (although not always) comes in the form of initial conditions, or values of y and $\dfrac{dy}{dt}$ when $t = 0$. Here are some worked examples, and exercises for you to do.

Example 16.7.1

From DO THIS NOW! 16.2 (i) $\dfrac{d^2 y}{dt^2} + 5\dfrac{dy}{dt} + 4y = 0$ has general solution

$$y = Ae^{-1t} + Be^{-4t} .$$

Suppose the initial conditions are $y = 0$ and $\dfrac{dy}{dt} = 1$ when $t = 0$. You can apply these as follows:

$y = 0$ when $t = 0$ gives $\qquad 0 = Ae^{-1\times 0} + Be^{-4\times 0} = A + B$.   (16.7)

$\dfrac{dy}{dt} = 1$ when $t = 0$ requires $\dfrac{dy}{dt} = -Ae^{-1t} - 4Be^{-4t}$, and putting in the condition gives

$$1 = -Ae^{-1\times 0} - 4Be^{-4\times 0} = -A - 4B .$$ (16.8)

These equations (16.7) and (16.8) can now be solved simultaneously for A and B, giving values $A = \dfrac{1}{3}$ and $B = -\dfrac{1}{3}$, and so the particular solution satisfying the initial conditions is:

$$y = \frac{1}{3}e^{-1t} - \frac{1}{3}e^{-4t}$$

A quick check via a CAS solver, and a plot, are given in Figure 16.2 to verify this is correct and looks right in terms of the system and its initial conditions.

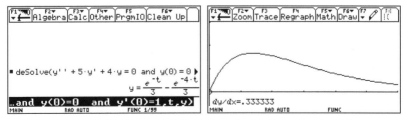

Figure 16.2 Checking an O.D.E. solution using a built-in solver in CAS.

Example 16.7.2

From DO THIS NOW! 16.2 (ii) $\dfrac{d^2 y}{dt^2} + 4\dfrac{dy}{dt} + 4y = 0$ has general solution

$$y = (A + Bt)e^{-2t}.$$

Suppose the boundary conditions are $y = 0$ when $t = 0$ and $y = 1$ when $t = 1$ (in this case they are not both initial conditions but the essential method is the same), and you can apply these as follows:

$y = 0$ when $t = 0$ gives $\qquad 0 = (A + B \times 0)e^{-2\times 0} = A$. $\qquad$ (16.9)

$y = 1$ when $t = 1$ gives $\qquad 1 = (A + B \times 1)e^{-2\times 1}$ or $e^2 = A + B$. $\qquad$ (16.10)

(16.9) and (16.10) give values $A = 0$ and $B = e^2 \approx 7.39$, and so the particular solution satisfying these initial conditions is

$$y = 7.39te^{-2t}.$$

Once more a quick plot can be used (Figure 16.3) to verify that this looks right in terms of the system and its initial conditions.

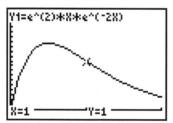

Figure 16.3 Using a plot of a solution to see if it looks sensible.

Example 16.7.3

As a final example, take DO THIS NOW! 16.2 (iii) $\dfrac{d^2 y}{dt^2} + 4\dfrac{dy}{dt} + 5y = 0$. This has

general solution $\qquad\qquad y = e^{-2t}(A\cos(1t) + B\sin(1t))$.

Suppose the initial conditions are $y = 0$ and $\dfrac{dy}{dt} = 1$ when $t = 0$. You can apply these as follows:

$y = 0$ when $t = 0$ gives $\qquad 0 = e^{-2\times 0}(A\cos(1\times 0) + B\sin(1\times 0)) = A \qquad$ (16.11)

and so $A = 0$ and there is no cosine term in the answer.

$\dfrac{dy}{dt} = 1$ when $t = 0$ requires $\dfrac{dy}{dt} = \dfrac{d}{dt}(e^{-2t} B\sin(1t)) = -2e^{-2t} B\sin(1t) + e^{-2t} B\cos(1t)$,

and putting in the condition gives

$$1 = -2e^{-2 \times 0} B \sin(1 \times 0) + e^{-2 \times 0} B \cos(1 \times 0) = B \qquad (16.12)$$

(16.11) and (16.12) then give values $A = 0$ and $B = 1$, and so the particular solution satisfying these initial conditions is

$$y = e^{-2t} \sin(1t))$$

Yet again a quick plot can be used (Figure 16.4) to verify that this looks right in terms of the system and its initial conditions.

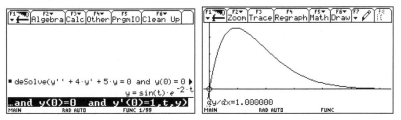

Figure 16.4 Verifying and plotting a solution using CAS.

DO THIS NOW! 16.3

Repeat the examples above to find the values of A and B for the following boundary conditions:

For DO THIS NOW 16.2 (i), use $y = 1$ and $\dfrac{dy}{dt} = 1$ when $t = 0$.

For DO THIS NOW 16.2 (ii), use $y = 0$ and $\dfrac{dy}{dt} = 1$ when $t = 0$.

For DO THIS NOW 16.2 (iii), use $y = 0$ when $t = 0$, and $y = 1$ when $t = \dfrac{\pi}{2}$.

In each case plot your resulting solution to see how it behaves and to check that the boundary conditions are indeed satisfied.

DO THIS NOW! 16.4

Here is an exercise to let you explore these ideas a little further.

Solve the simplified standard O.D.E. $\dfrac{d^2 y}{dt^2} + b\dfrac{dy}{dt} + cy = 0$ for an under-damped case, say where $b = 2$ and $c = 2$. What happens if you fix the spring term c at a value of 2, and reduce the damping term b until you eventually remove it altogether? What does the O.D.E. look like in this case? What is its solution? Plot the solution and comment on it (use as your initial conditions $y(0) = 0$ and $\dfrac{dy}{dt}(0) = 1$).

Hint: Exponential part of the solution disappears and leaves undamped oscillations. Eventually $x(t) = \sin(t)$. Not unexpected since you have removed the damping term.

16.8 FINDING THE P.I. - THE EFFECT OF FORCING CHANGE.

So far in finding the CF you have only been looking at unforced motion - that is the case where the RHS of the O.D.E. is 0. Once the RHS is no longer 0, an external "force" influences the motion, and an extra term is needed in the solution to take account of that force, called the Particular Integral (PI) or steady state term. This can be found by a variety of methods, one of which includes intelligent guesswork, using the observation that the steady state of a system will be driven by, and will therefore mirror in some sense, the external force driving it.

Example 16.8.1

Find the general solution of $\dfrac{d^2 y}{dt^2} + 4\dfrac{dy}{dt} + 3y = 6$.   (16.13)

The overall solution is $y(t) = \text{CF} + \text{PI}$.

For the CF we solve the auxiliary equation $m^2 + 4m + 3 = 0$ as above. This gives:

$$\text{CF} = y(t) = Ae^{-3t} + Be^{-t} .   (16.14)$$

Now the PI term is to take account of the external force – and referring back to (16.3) this force is on the RHS of the equation. In this case RHS = 6 is equivalent to a constant external force, so it reasonable to guess that the steady state response to that will also be constant. Thus we guess that the Particular Integral (PI) is $y(t) = k$ (a constant) and substituting this into (16.13) we find that PI = $k = 2$.

(Note that no further arbitrary constants are coming into this. We already have the number of arbitrary constants we need from the CF part of the solution.)

Thus the general solution is given by

$$y(t) = \text{CF} + \text{PI} = Ae^{-3t} + Be^{-t} + 2 .$$

DO THIS NOW! 16.5

(i) Find the general solution of $\dfrac{d^2 y}{dt^2} + 4\dfrac{dy}{dt} + 3y = 9$.

Hint: for the P.I., guess $y(t) = K$ and find K by substitution in the equation.

Find and plot the solution for the case where $y = 1$ and $\dfrac{dy}{dt} = 1$ when $t = 0$.

(ii) Find the general solution of $\dfrac{d^2 y}{dt^2} + 4\dfrac{dy}{dt} + 3y = e^{-2t}$.

Hint: for the P.I., guess $y(t) = Ke^{-2t}$ and find K by substitution in the equation.

Find and plot the solution for the case where $y = 1$ and $\dfrac{dy}{dt} = 1$ when $t = 0$.

You can see that the general principle is that you base your guess for the PI on the shape of the RHS, and substitute it into the equation to see if it works. We shall not dwell on this further here, but will move on to alternative, practical approaches, both technological and analytical.

16.9 TECHNOLOGICAL SOLVERS
16.9.1 Types of solver

There are many software packages to solve O.D.E.s, both analytically (if it is possible) and numerically. These include CAS packages both hand-held and desktop such as Derive, TI-Interactive, Mathcad and Mathematica, and numerical approaches implemented through a spreadsheet or programming language.

DO THIS NOW! 16.6

Find out now what CAS packages you have available to you which will solve second order O.D.E.s, and use them to solve the equations in DO THIS NOW 16.5! and earlier exercises.

You will find that these solvers will provide solutions to all the standard problems, but in many real problems, engineers must use more flexible numerical approaches. You have the option to explore numerical approaches further in the End-of-Chapter exercises.

16.9.2 Exploring the effect of varying the parameters

Some software packages allow the exploration of families of solution of an O.D.E by varying the parameters, for instance a, b, c, and constants in $f(t)$ in equation 16.1. This ability to "play" with parameters can be particularly helpful, for instance in trying to understand and optimize system behaviour. The process can be supported in some cases by the use of sliders, as illustrated in the TI-Interactive extract in Figure 16.5.

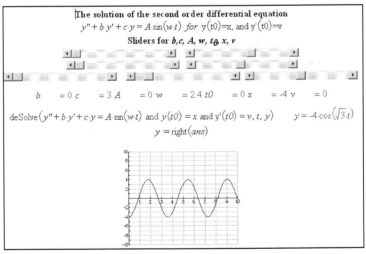

Figure 16.5 Exploring the effects of varying the parameters in TI-Interactive.

DO THIS NOW! 16.7

Find out now if your CAS system supports the use of sliders, and use what you learn to revisit DO THIS NOW 16.4! using any built-in O.D.E. solver, and varying b using a slider if that is possible.

DO THIS NOW! 16.8

The suspension system introduced in section 16.3 and shown in Figure 16.1, exhibits behaviour which is modelled by the O.D.E.

$$\frac{d^2y}{dt^2}+\frac{k}{M}\frac{dy}{dt}+\frac{\lambda}{M}y=F(t)\ .$$

One particular system with appropriate values of the parameters is modelled by:

$$\frac{d^2y}{dt^2}+0.6\frac{dy}{dt}+3.3y=0\ .$$

By finding the CF, write down the general solution of this equation.

The initial conditions are $y = 0.5$ and $\frac{dy}{dt} = 2$ when $t = 0$. Show that the solution in this case is

$$y=e^{-0.3t}(0.5\cos(1.79t)+1.2\sin(1.79t))\ .$$

This solution that results is displayed graphically in Figure 16.6:

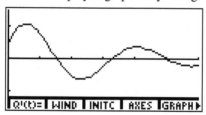

Figure 16.6 Displaying the solution graphically is a powerful confirmation.

Set up your CAS solver to confirm the solution to this equation.

By varying the appropriate parameter, by means of sliders if possible, describe briefly with the aid of sketches what happens to the solution and system when:

(i) The damping constant $\frac{k}{M}$ of the system increases or decreases from 0.6;

(ii) The spring constant $\frac{\lambda}{M}$ of the system increases or decreases from 3.3.

16.10 SOME FINAL EXAMPLES

We can pull together many of the threads of this chapter through a final example.

Example 16.10.1

Solve the O.D.E. $\frac{d^2y}{dt^2}+0.1\frac{dy}{dt}+y=e^{-0.5t}$ subject to conditions $y(0)=\frac{dy}{dt}(0)=0$.

DO THIS NOW! 16.9

(a) Show that the roots of the auxiliary equation are $m = -0.05000 \pm j\, 0.99875$.
(Figure 16.7 gives a check by graphic calculator!)

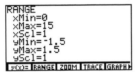

Figure 16.7 Checking quadratic roots by graphical calculator.

(b) Show that the CF is $CF = e^{-0.05t}(A\cos(0.99875t) + B\sin(0.99875t))$.

(c) By guessing $y = ke^{-0.5t}$ show that $PI = 0.83333\, e^{-0.5t}$.

(d) Hence write down the solution $y(t) = CF + PI$.

(e) By using the boundary conditions show that $A = -0.83333$ and

$B = 0.37547$.

(e) Hence write down the overall solution.

(f) Check this solution using your technological solver.

(g) To see if this is reasonable, plot this solution (see Figure 16.8):

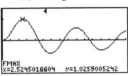

Figure 16.8 Plot you solution to see if it looks sensible.

The graph looks reasonable; it displays oscillations and damping, and the boundary conditions appear to be satisfied,

We should still ask is the detail of the solution valid? The numbers involved are excruciating. Most books have "nice" values but engineering problems are not like that, and there are many possible sources of error: rounding, carelessness, etc. We need some more convincing.

DO THIS NOW! 16.10

Check your answer to DO THIS NOW! 16.9 using your local CAS package(s). Get a plot of the solution. Do you get something like Figure 16.9, produced in Derive?

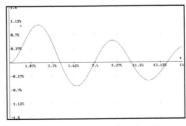

Figure 16.9 Checking again using Derive.

END OF CHAPTER 16 MIXED EXERCISES – DO ALL OF THESE NOW!

1. Solve the differential equation

$$\frac{d^2 y}{dt^2} + n^2 y = 0$$

where t is the time, y is the displacement of a particle along a line, and n is a constant. Show that the motion of the particle is periodic and of period $\frac{2\pi}{n}$.

2. Solve the ordinary differential equation

$$\frac{d^2 y}{dt^2} + k\frac{dy}{dt} + 6y = 2$$

where $k = 5$ and the initial conditions are $y(0) = 0$ and $\frac{dy}{dt}(0) = 0$.

Identify the transient and steady-state terms in the solution, state whether the system described by the equation is over-damped or under-damped.

What value of k must be taken to ensure critical damping?

3. The equation of motion of a person on a bouncing toy which consists of a seat on a spring bouncing in a vertical direction is given by the differential equation:

$$m\frac{d^2 y}{dt^2} + 9\frac{dy}{dt} + 120y = \sin(2t) ,$$

where y is the distance measured vertically upwards from the position of the seat when the person first sits on it, m is the mass of the person and t represents time.

In equilibrium, when the person sits on the seat at rest, the value of y is 0.

In the following problems suppose that the person is initially at rest at $y = 0.2$, obtain a graph of the motion of the device for a person of mass 20.

Conjecture how does this graph change if the person has mass 10, 30, 40, 50? Plot the graphs for these cases. Do the graphs confirm your conjecture(s)?

4. Consider the equation $\frac{d^2 y}{dt^2} + p\frac{dy}{dt} + qy = R\sin(\omega t)$.

Set up your CAS system to provide a solution to this equation for the case where $R = 0, p = 1, q = 4$, with initial conditions $y(0) = 0$ and $\frac{dy}{dt}(0) = 1$.

You should find the solution is $y(t) = 0.516e^{-0.5t}\sin(1.936t)$. A suitable plot of this function is a straightforward two-dimensional graph of y against t.

Question 5 continues with this problem, but looking at alternative ways of representing the overall behaviour of the system.

5. Take the equation of question 4 and set up your CAS system to provide a solution for the case where $R = 0$ and $q = 4$, with initial conditions $y(0) = 0$ and $y'(0) = 1$, but this time leaving damping p unspecified.

Thus your system should give y as a function of two variables

$$y(t, p) = \frac{2e^{-0.5pt}\sin(0.5t\sqrt{16-p^2})}{\sqrt{16-p^2}}.$$

(We produced this using the DSOLVE2 function in Derive.)

Now consider how you might usefully plot this to represent the information contained in the formula, and to explore and explain the effect of varying the parameter p.

One possibility is to use 3D plotting facilities, plotting $y(t, p)$ on the vertical axis against t and p on the two horizontal axes, producing a graph like that from Derive in Figure 16.10.

Interpret what this graphical object is telling you. If you follow any line along the t direction you will see a time plot for a particular value of p. If you look how these plots change as you move across the surface in the p direction you will see the system making the transition from under- to over-damping. Study the graph and make sense of it in your own words.

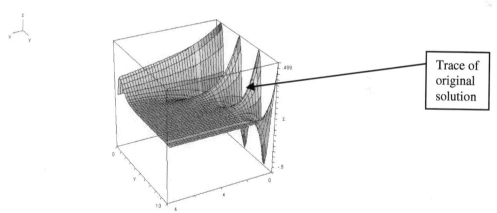

Figure 16.10 Different ways of "seeing" your solutions can help you understand and check them.

6. We mentioned that you could extend a numerical method such as Euler's method to second order differential equations. In this problem you will have a brief opportunity to explore that – although this will only be a sketch treatment. You might like to see if you can fill in the details yourself, and make it work. We give you some details and set it as a kind of detective trail – can you decode the spreadsheet, see what we have done, and make a similar spreadsheet work for yourself?

Consider then the equation $\dfrac{d^2y}{dt^2} + k_1\dfrac{dy}{dt} + k_2y = 0$.

The general idea is to represent this as two first order equations, and then to solve them together using Euler's method as explained in Chapter 15.

Thus we define $x_1 = y$ and $x_2 = \dfrac{dy}{dt}$, and in that case the single second order O.D.E. becomes two first order O.D.E.s:

$$\frac{dx_1}{dt} = x_2$$

$$\frac{dx_2}{dt} = -k_2x_1 - k_1x_2$$

These equations are expressed in numerical form for Euler's method as

$$x_1(new) = x_1(old) + hx_2$$
$$x_2(new) = x_2(old) + h(-k_2x_1 - k_1x_2)$$

We take $k_1 = 0.6$, $k_2 = 41.5$, $y(0) = -0.46$ and $y'(0) = 4.14$. (This was real data observed from a bungee jump!) We use a time step of 0.01.

The spreadsheet in Figure 16.11 shows our implementation of this procedure, with the plot being x_1 (or y) against t. Can you decode and reconstruct?

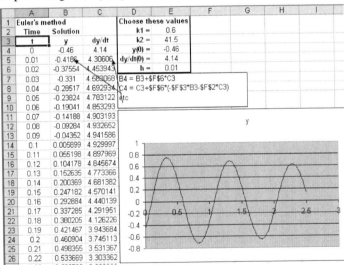

Figure 16.11 Extending Euler's method to second order equations: can you decode?

Good luck with this. Remember that you may have to reduce the time step until you are convinced that you are getting an accurate answer. You can always check your answers using another method, such as a built-in CAS tool.

17

Laplace Transforms And Ordinary Differential Equations

"It's a kind of magic. There can be only one." Queen

17.1 THE USEFULNESS OF THE LAPLACE TRANSFORM

The Laplace transform is one of several transforms you may meet in engineering. It is perhaps the one most commonly used. It takes problems expressed in terms of time t (which may appear as ordinary differential equations) and by using a mathematical transformation, re-expresses them in terms of another variable s, in such a way that the problem becomes easier to solve. The method is so powerful because what the transform does is to turn a *differential* equation involving t into merely an *algebraic* equation involving the new variable s. Thus, while it may look magical and laborious, it gives a simpler approach to complex problems.

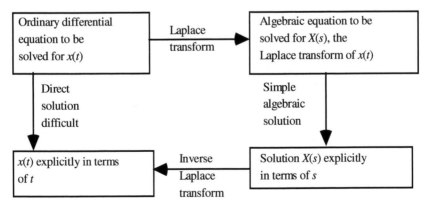

Figure 17.1 A diagrammatic view of how the Laplace transform works.

You will find that the Laplace transform is much more than just a way of solving ODEs, being also a powerful language to describe the behaviour of control systems, electronic circuits, communication systems, and other engineering devices.

To become familiar with this technique, keep on **doing the exercises and activities**.

17.2 WHAT IS THE LAPLACE TRANSFORM?
17.2.1 Definition of the Laplace transform

The transform from time domain to s domain is carried out using the formula:

$$X(s) = \int_0^\infty x(t)e^{-st}dt \ . \tag{17.1}$$

where $x(t)$ (little x) is the function in question, and $X(s)$ (capital X, sometimes written $L(x(t))$) is its Laplace transform. Points to note are:

- t disappears when the integral is done, which is why $X(s)$ is a function of s;

- s behaves something like a frequency. It is in fact a complex number. It is best to put its meaning and significance on hold until you are used to seeing it in action;

- this integral is messy to do and in some cases difficult as well, and so engineers use previously built-up tables of Laplace transforms to allow them to move easily between t and s.

In relation to this last point, we shall not derive the transforms of every function from the definition, but here is one simple example just to see what is involved.

Example 17.2.1 The Laplace transform of a constant.

Find the Laplace transform of the function $x(t) = k$ where (k is a constant).

Using the definition (17.1) the Laplace transform is:

$$X(s) = \int_0^\infty x(t)e^{-st}dt$$

$$= \int_0^\infty ke^{-st}dt = k\int_0^\infty e^{-st}dt$$

$$= k\left[-\frac{1}{s}e^{-st}\right]_0^\infty$$

$$= k(0) - k(-\frac{1}{s}) = \frac{k}{s}$$

Note we have assumed that everything goes to 0 at the top, infinite limit. This turns out to be a safe assumption with the functions we need to use.

Thus if $x(t) = k$ (constant) then its Laplace transform is

$$X(s) = L(x(t)) = \frac{k}{s} \ . \tag{17.2}$$

17.2.2 Building up a table of Laplace transforms

You can build up a table of the Laplace transforms of all the common functions, by going through a process similar to that in Example 17.2.1. Table 17.1 below shows some of the most commonly occurring ones:

$x(t)$	$X(s)$
k	$\dfrac{k}{s}$
$ke^{-\alpha t}$	$\dfrac{k}{s+\alpha}$
$k\sin(\omega t)$	$\dfrac{k\omega}{s^2+\omega^2}$
$k\cos(\omega t)$	$\dfrac{ks}{s^2+\omega^2}$
$k(1-e^{-\alpha t})$	$\dfrac{k\alpha}{s(s+\alpha)}$
$\dfrac{dx}{dt}$	$sX(s)-x(0)$

Table 17.1 Common Laplace transforms.

Fuller tables follow at the end of the chapter, but Table 17.1 will suffice for now.

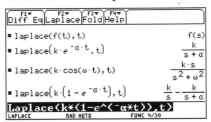

Figure 17.2 Laplace transforms can also be handled on some CAS.

Some CAS technology has the Laplace transform built in, as shown for instance in Figure 17.2. You will find that the amount of manipulation you need to do to use the method depends on the depth of your tables or the power of your technology.

DO THIS NOW! 17.1

1. Derive by hand at least one of the other formulae in Table 17.1 from the definition of the Laplace Transform in equation (17.1) (the next easiest is perhaps that for $x(t) = ke^{-\alpha t}$).

2. Find out now whether your CAS can handle Laplace transforms and, if so, use it to check some of the entries in Table 17.1.

17.2.3 The Laplace transform of $\dfrac{dx}{dt}$.

The last entry in Table 17.1 shows why the Laplace transform is so useful in solving differential equations. You can see that it turns differentiation of x with respect to t into simple multiplication of X by s (give or take an $x(0)$ term, that is one involving the initial value of x). This is an important result and here is its derivation:

$$L(\frac{dx}{dt}) = \int_0^\infty \frac{dx}{dt} e^{-st} dt = \left[x(t)e^{-st} \right]_0^\infty - \int_0^\infty x(t)(-se^{-st}) dt \quad \text{(using integration by parts)}$$

$$= -x(0) + s \int_0^\infty x(t)e^{-st} dt = sX(s) - x(0) .$$

Do not worry if this algebra seems difficult, but make sure you note the result:

• this is just s times $X(s)$, the Laplace transform of $x(t)$, together with an adjustment term involving $x(0)$;

• we have assumed again that at the infinite limit everything goes to 0.

17.3 THE LAPLACE TRANSFORM IN ACTION: FIRST ORDER ODES

In this section you will meet examples that show how the result of section 17.2.3 allows the solution of simple first order linear ODEs, of the type you have previously met in Chapters 10 and 15. The examples will follow the process through from start to finish, and although they are very simple cases, they will contain all the essential points of the method.

Example 17.3.1

Solve the first order linear ODE $\dfrac{dx}{dt} + x = 1$, subject to the initial condition $x(0) = 0$.

The Laplace transform is used like this:

Step 1: Take the Laplace transform of the *whole* equation $L\left[\dfrac{dx}{dt} + x = 1\right]$.

This leads to $\qquad L\left[\dfrac{dx}{dt}\right] + L[x] = L[1]$.

Each of these terms can be found from Table 17.1 to give

$$sX(s) - x(0) + X(s) = \frac{1}{s} .$$

Step 2: Put in the initial condition $x(0) = 0$:

$$sX(s) - 0 + X(s) = \frac{1}{s} .$$

Step 3: Solve for $X(s)$: $\qquad (s+1)X(s) = \frac{1}{s}$,

which may be re-arranged $X(s) = \dfrac{1}{s(s+1)}$.

Step 4: Take the inverse Laplace transform.

Use Table 17.1 from right to left choosing the correct entry in terms of where s is. That is, pick the formula $\dfrac{k\alpha}{s(s+\alpha)}$ and choose $k = 1$ and $\alpha = 1$.

This gives the solution $x(t) = 1 - e^{-t}$

Note that:

- this solution could also have been obtained by separation of variables;

- the arbitrary constant is implicitly found as the initial condition is built in;

- CAS technology can deal with both the inverse Laplace transform, and indeed direct solution of the ODE, as shown for instance in Figure 17.3.

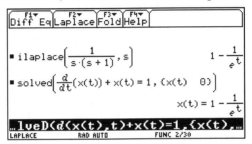

Figure 17.3 Finding the inverse Laplace transform using CAS.

DO THIS NOW! 17.2

Make sure you understand Example 17.3.1, then, using the same method, solve the following equations (a) using Laplace transforms, and (b) checking by either using CAS technology or separation of variables:

1. $\dfrac{dx}{dt} + 2x = 2$ subject to $x(0) = 0$ (Sol. $x(t) = 1 - e^{-2t}$).

2. $\dfrac{dx}{dt} + 1x = 2$ subject to $x(0) = 0$ (Sol. $x(t) = 2(1 - e^{-t})$).

3. $\dfrac{dx}{dt} + 5x = 10$ subject to $x(0) = 0$ (Sol. $x(t) = 2(1 - e^{-5t})$).

You have now seen how the Laplace transform can be used to solve ODEs. More complicated cases can be dealt with in essentially the same way, but with a little more algebra. There is often more than one way through the algebra of a problem; for example sometimes you could use partial fractions or completing the square, but this extra algebra can often be avoided if your tables are comprehensive enough.

We cannot hope to cover every possible case, but here is another example to show a different problem. - you have to make sure you know your way around your tables, and use them well, combining this with some algebraic imagination and technological support. The tables referred to are those provided at the end of this chapter. Follow the next example and plot out the solution to see what it looks like.

Example 17.3.2.

Solve $$\frac{dx}{dt} + 2x = e^{-t} \quad \text{subject to } x(0) = 0.$$

Step 1: Using tables, take Laplace transforms of both sides:

$$[sX(s) - x(0)] + 2X(s) = \frac{1}{s+1}.$$

Step 2: Using $x(0) = 0$:

$$(s+2)X(s) = \frac{1}{s+1}.$$

Step 3: Thus

$$X(s) = \frac{1}{(s+1)(s+2)}.$$

Step 4: To invert, look at the tables. The correct form is $\dfrac{k}{(s+\alpha)(s+\beta)}$ with $k = 1$,

$\alpha = 1$ and $\beta = 2$.

Thus the inverse is

$$x(t) = \frac{k}{(\beta - \alpha)}(e^{-\alpha t} - e^{-\beta t})$$

$$= \frac{1}{(2-1)}(e^{-1t} - e^{-2t}) = e^{-1t} - e^{-2t}.$$

You can confirm this solution for instance by CAS, as in Figure 17.4, although the CAS used there (TI Voyage 200) presents the solution in a slightly different form.

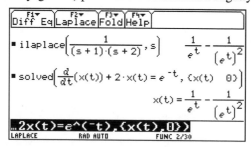

Figure 17.4 Confirming your solution using CAS.

DO THIS NOW! 17.3

Make sure you understand the examples above. Then, using the same method, solve the following first order linear ODEs using the Laplace transform, and check your answers otherwise:

1. $\dfrac{dx}{dt} + 3x = 5e^{-t}$ \qquad subject to $x(0) = 0$ \qquad (Sol. $\dfrac{5}{2}(e^{-t} - e^{-3t})$)

2. $\dfrac{dx}{dt} + 3x = \sin(t)$ \qquad subject to $x(0) = 0$ \qquad (Sol. $\dfrac{1}{10}(e^{-3t} - \cos(t) + 3\sin(t))$)

17.4 USING THE LAPLACE TRANSFORM WITH SECOND ORDER ODEs

You can solve second order ODEs using basically the same steps as for first order ODEs. You will find that the main extra complication occurs because there will be a quadratic term, corresponding to the quadratic auxiliary equation. As you have seen in Chapter 16, the nature of the solution varies according to whether this quadratic has real roots or not. The solution therefore depends on "$(b^2 - 4ac)$". It is not surprising that this issue turns up in Laplace transform approaches, as these are the same ODEs we are solving!

The steps are the same, but extra complication will occur when you look at the tables to try to find your inverse. If your tables are comprehensive enough then each form will be there; if not then you may have to do some algebraic pre-processing such as partial fractions or completing the square. Another difference is that two initial conditions are needed for second order equations. We reiterate that we cannot cover every type of example that will turn up – you will need your tables, technology and algebraic skills! The following examples illustrate a typical range of problems.

Example 17.4.1 A second order ODE.

Solve $\qquad\qquad\qquad \dfrac{d^2x}{dt^2} + 9x = 1 \qquad$ subject to $x(0) = x'(0) = 0$.

We shall now move to using Table 17.2 at the end of this chapter; it includes the results from Table 17.1, but is more comprehensive.

Step 1: Using Table 17.2, take Laplace transforms of both sides of the equation (you will need to use the transform of a second derivative now):

$$[s^2 X(s) - sx(0) - x'(0)] + 9X(s) = \frac{1}{s}$$

Step 2: Using $x(0) = x'(0) = 0$: $\qquad (s^2 + 9)X(s) = \dfrac{1}{s}$.

Step 3: Solving for $X(s)$: $\qquad\qquad X(s) = \dfrac{1}{s(s^2 + 9)}$.

Step 4: To invert, look at Table 17.2, and find the correct form, which is

$$\frac{k}{(s+\alpha)(s^2 + \omega^2)} \quad \text{with } k = 1,\ \alpha = 0 \text{ and } \omega = 3.$$

Thus the inverse is: $\qquad x(t) = \dfrac{k}{(\alpha^2 + \omega^2)}(e^{-\alpha} - \cos(\omega t) + \dfrac{\alpha}{\omega}\sin(\omega t))$

$$= \frac{1}{(0^2 + 3^2)}(e^{-0t} - \cos(3t) + \frac{0}{3}\sin(3t))$$

$$= \frac{1}{9}(1 - \cos(3t))\ .$$

You have more than one means to check the outcome here. You could use CAS as illustrated in Figure 17.5, either as a table substitute, or indeed to solve the ODE directly! Alternatively, you could substitute the solution into the ODE and see if it works, checking that the initial conditions are satisfied too, of course.

Figure 17.5 Checking again using CAS.

Here is another example.

Example 17.4.2 Another second order ODE

Solve
$$\frac{d^2 x}{dt^2} + 4\frac{dx}{dt} + 3x = 0 \qquad \text{subject to } x(0) = 0 \text{ and } x'(0) = 1.$$

Step 1: Using Table 17.2 again, take Laplace transforms of both sides:

$$[s^2 X(s) - sx(0) - x'(0)] + 4[sX(s) - x(0)] + 3X(s) = 0 .$$

Step 2: Using $x(0) = 0$ and $x'(0) = 1$:

$$(s^2 + 4s + 3)X(s) - 1 = 0 .$$

Step 3: Thus $\quad X(s) = \dfrac{1}{(s^2 + 4s + 3)} .$

Step 4: Invert this Laplace transform. First, check whether the auxiliary quadratic $m^2 + 4m + 3 = 0$ has real roots using $b^2 - 4ac = 16 - 12 = 4 > 0$. This is positive so it has real factors $(x - root1)(x - root2)$ where *root1* and *root2* are the roots of the quadratic. The roots are –1 and –3, and thus the factors in this case are $(x + 1)(x + 3)$, and the transform can be rewritten

$$X(s) = \frac{1}{(s+1)(s+3)} .$$

Looking at Table 7.2, the correct standard form is $\dfrac{k}{(s+\alpha)(s+\beta)}$, choosing $k = 1$, $\alpha = 1$ and $\beta = 3$, and thus the tables say the inverse is

$$x(t) = k[\frac{1}{(\beta - \alpha)}(e^{-\alpha t} - e^{-\beta t})] = 1[\frac{1}{(3-1)}(e^{-1t} - e^{-3t})] = \frac{1}{2}(e^{-t} - e^{-3t}) .$$

DO THIS NOW! 17.4

Find the solution of this ODE via the complementary function. Do you get the same result? (You should!)

You can check the result by substituting into the original ODE, although this becomes more complicated as the ODE does so; or you can use CAS technology as in Figure 17.6; or you can use an alternative method as in **DO THIS NOW 17.4!**

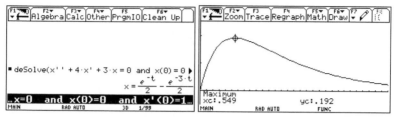

Figure 17.6 Always think about whether you answer is right, and check by other means if you can.

Example 17.4.3 A more complicated second order ODE

Solve $\dfrac{d^2x}{dt^2}+4\dfrac{dx}{dt}+3x=1$ subject to $x(0) = 0$ and $x'(0) = 0$.

Step 1: Using Table 17.2 again, take Laplace transforms of both sides:

$$[s^2X(s)-sx(0)-x'(0)]+4[sX(s)-x(0)]+3X(s)=\frac{1}{s} \ .$$

Step 2: Using $x(0) = x'(0) = 0$:

$$(s^2+4s+3)X(s)=\frac{1}{s} \ .$$

Step 3: Thus $X(s)=\dfrac{1}{s(s^2+4s+3)},$

and factorising the quadratic as in Example 17.4.2:

$$X(s)=\frac{1}{s(s+1)(s+3)} \ .$$

Step 4: To invert, look at Table 17.2: the correct standard form is

$$\frac{k}{(s+\alpha)(s+\beta)(s+\gamma)} \ \text{with } k = 1, \ \alpha = 0, \ \beta = 1 \text{ and } \gamma = 3.$$

From Table 17.2 then, the inverse is:

$$x(t)=k[\frac{1}{(\beta-\alpha)(\gamma-\alpha)}e^{-\alpha t}+\frac{1}{(\alpha-\beta)(\gamma-\beta)}e^{-\beta t}+\frac{1}{(\alpha-\gamma)(\beta-\gamma)}e^{-\gamma t}]$$

$$=1[\frac{1}{(1-0)(3-0)}e^{-0t}+\frac{1}{(0-1)(3-1)}e^{-1t}+\frac{1}{(0-3)(1-3)}e^{-3t}]$$

$$=\frac{1}{3}-\frac{1}{2}e^{-1t}+\frac{1}{6}e^{-3t}$$

Once more you have the usual range of checks (see Figure 17.7).

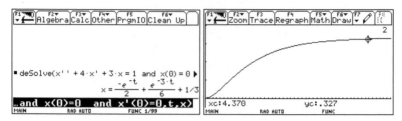

Figure 17.7 ... and checking again ...

Example 17.4.4 An under-damped, second order ODE

Solve $\dfrac{d^2x}{dt^2} + 2\dfrac{dx}{dt} + 26x = 0$ subject to $x(0) = 0$ and $x'(0) = 1$.

Step 1: Using Table 17.2 to take Laplace transforms of both sides:

$$[s^2 X(s) - sx(0) - x'(0)] + 2[sX(s) - x(0)] + 26X(s) = 0 \ .$$

Step 2: Using $x(0) = 0$ and $x'(0) = 1$:

$$(s^2 + 2s + 26)X(s) - 1 = 0 \ .$$

Step 3: Thus $X(s) = \dfrac{1}{(s^2 + 2s + 26)}$.

Step 4: To invert, we must begin by checking whether the auxiliary quadratic has real factors. In this case $b^2 - 4ac = 4 - 104 < 0$, so there are no real factors, and we have an under-damped system. In this case the best way to proceed is to complete the square on the quadratic expression:

$$s^2 + 2s + 26 = (s+1)^2 + 25 = (s+1)^2 + 5^2 \ .$$

The transform can then be written:

$$X(s) = \dfrac{1}{((s+1)^2 + 5^2)} \ .$$

This form appears in Table 17.2 as $\dfrac{k}{((s+\beta)^2 + \omega^2)}$, and we must clearly choose $k = 1$, $\beta = 1$ and $\omega^2 = 5$. Table 17.2 then says the inverse is

$$x(t) = \dfrac{k}{\omega} e^{-\beta t} \sin(\omega t) = \dfrac{1}{5} e^{-1t} \sin(5t) \ .$$

DO THIS NOW! 17.5

Find the solution of this under-damped ODE via the complementary function. Do you get the same result? (You should!)

Again you have the range of ways of checking what you have done, including using CAS technology, as illustrated in Figure 17.8.

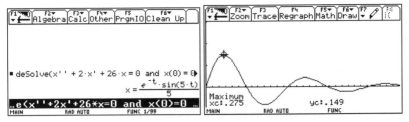

Figure 17.8 Checking an under-damped case.

DO THIS NOW! 17.6

Make sure you understand the examples above. Then, using the same method, solve the following equations:

1. $\dfrac{d^2x}{dt^2} + 25x = e^{-t}$, subject to $x(0) = x'(0) = 0$ (Sol. $\dfrac{1}{26}(e^{-t} - \cos(5t) + \dfrac{1}{5}\sin(5t))$).

2. $\dfrac{d^2x}{dt^2} + 3\dfrac{dx}{dt} + 2x = 2$, subject to $x(0) = x'(0) = 0$ (Sol. $1 - 2e^{-t} + e^{-2t}$).

3. $\dfrac{d^2x}{dt^2} + 5\dfrac{dx}{dt} + 6x = e^{-t}$, subject to $x(0) = x'(0) = 0$ (Sol. $\dfrac{1}{2}(e^{-t} - 2e^{-2t} + e^{-3t})$).

4. $\dfrac{d^2x}{dt^2} + 5\dfrac{dx}{dt} + 6x = 0$, subject to $x(0) = 0$ and $x'(0) = 4$ (Sol. $4(e^{-2t} - e^{-3t})$).

5. $\dfrac{d^2x}{dt^2} + 5\dfrac{dx}{dt} + 6x = 0$, subject to $x(0) = 4$ and $x'(0) = 0$ (Sol. $12e^{-2t} - 8e^{-3t}$).

17.5 A FINAL SPECIAL CASE – RESONANCE

This final example raises a possibility of which every engineer should be aware – the phenomenon of resonance. Resonance is associated with vibrations. Second order vibrating engineering systems have a natural frequency at which they will vibrate, given by the complementary function part of the solution as found in chapter 16. If an external force oscillates at the same natural frequency then the vibrations in a system can grow with catastrophic results.

One example of resonance occurred when the Broughton Bridge near Manchester, England collapsed in 1831 as a result of soldiers marching over the bridge in step at just the right (or wrong!) marching rate. Another well-known example involves shattering wine glasses with the amplified voices of singers.

Perhaps one of the most famous examples of the disastrous effects of resonance was the Tacoma Narrows bridge collapse in 1940, where the wind blew at the right speed and from the right direction to set up natural frequency vibrations which grew and led to the collapse of the bridge.

Resonance can occur in *any* vibrating system, including bridges, electrical circuits, and the human body. It is important in any system design that this phenomenon is recognized and taken into account or the results can be disastrous. Here is an example which illustrates how resonance arises mathematically.

Example 17.5.1 A potentially resonant vibrating system

You may find the algebra in this example particularly demanding, but even if that causes you a problem, make sure you understand the *meaning* of the solution!

Solve $\qquad \dfrac{d^2x}{dt^2} + 4x = \sin(\Omega t)$ subject to $x(0) = x'(0) = 0$.

Note the external force on the RHS vibrates at a frequency Ω.

Step 1: As before, use Table 17.2 to take Laplace transforms of both sides:

$$[s^2X(s) - sx(0) - x'(0)] + 4X(s) = \frac{\Omega}{s^2 + \Omega^2} \ .$$

Step 2: Using $x(0) = x'(0) = 0$: $\qquad (s^2 + 4)X(s) = \dfrac{\Omega}{s^2 + \Omega^2} \ .$

Step 3: Thus $\qquad\qquad\qquad X(s) = \dfrac{\Omega}{(s^2 + 4)(s^2 + \Omega^2)} \ .$

Step 4: Now when it comes to inverting, it is apparent from the tables that the correct form is different according to whether $\Omega^2 = 4$ or not.

Step 4 *The non-resonant case* $\Omega^2 \neq 4$.

For instance if $\Omega = 1$ the standard form is $\dfrac{k}{(s^2 + \omega^2)(s^2 + \Omega^2)}$

with $k = 1$, $\Omega = 1$ and $\omega = 2$.

Thus in this case the inverse is

$$x(t) = \frac{k}{(\omega^2 - \Omega^2)}(\frac{1}{\Omega}\sin(\Omega t) - \frac{1}{\omega}\sin(\omega t)) = \frac{1}{(2^2 - 1^2)}(\frac{1}{1}\sin(1t) - \frac{1}{2}\sin(2t))$$

$$= \frac{1}{3}(\sin(t) - \frac{1}{2}\sin(2t))$$

Figure 17.9 displays this solution, and it is clear that all is well – the solution does not grow uncontrollably, but remains at finite values.

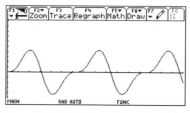

Figure 17.9 The solution of Example 17.5.1.

Step 5   *The resonant case* $\Omega^2 = 4$.

On the other hand if $\Omega^2 = 4$ then $\Omega = 2 = \omega$, and the vibrating external force on the RHS of the ODE is working at the same frequency as the natural vibration frequency of the system, so we must expect problems.

In detail, the standard form from Table 17.2 is $\dfrac{k}{(s^2 + \omega^2)^2}$ and we must choose $k = 1$, $\omega = \Omega = 2$, and in this case the inverse is

$$x(t) = \frac{k}{2\omega^3}(\sin(\omega t) - t\cos(\omega t)) = \frac{1}{2 \times 2^3}(\sin(2t) - t\cos(2t))$$

$$= \frac{1}{16}(\sin(2t) - t\cos(2t))$$

The plot of this in Figure 17.10 shows this solution (and the non-resonant solution) and shows how when $\Omega = 2$ the solution grows without bound as t increases. Thus the real physical phenomenon of resonance emerges naturally from the mathematical model we have constructed for a vibrating system in Chapter 16 and here. A squared quadratic factor tells you that something is going drastically wrong with your engineering system!

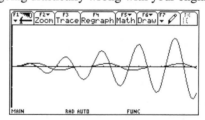

Figure 17.10 Graphical display of the resonance case.

DO THESE NOW! 17.7

Make sure you understand the examples above. Then, using the Laplace transform method, solve the following equations:

1. $\dfrac{d^2 x}{dt^2} + 25x = \sin(t)$, subject to $x(0) = x'(0) = 0$ (Sol. $\dfrac{1}{24}(\sin(t) - \dfrac{1}{5}\sin(5t))$).

2. $\dfrac{d^2 x}{dt^2} + x = \sin(t)$, subject to $x(0) = x'(0) = 0$ (Sol. $\dfrac{1}{2}(\sin(t) - t\cos(t))$.

Comment on the difference between these solutions.

END OF CHAPTER 17 EXERCISES – DO ALL THESE NOW!

1. Use the Laplace transform technique to find the solution of the ordinary differential equation

$$\frac{dy}{dt} + 5y = e^{-3t}$$

.

subject to the initial condition $y(0) = 0$. Calculate the value approached by $y(t)$ after a long time, and also the maximum value reached by $y(t)$, explaining how you reach your answers.

2. Using the Laplace transform technique, solve the ordinary differential equation

$$\frac{d^2y}{dt^2} + 3.6\frac{dy}{dt} + cy = 0$$

where $c = 2.88$ and the initial conditions are $y(0) = 0$ and $y'(0) = 1$.

Identify the transient and steady state terms in the solution and, giving reasons, state whether the system described by the equation is over-damped or under-damped.

What value of c must be taken to ensure critical damping?

3. (Beware! Complicated algebraic tricks required if you do this by pen and paper!) An engineering system is described by the ordinary differential equation

$$\frac{d^2y}{dt^2} + 1.44y = 5.1e^{-0.1t}$$

The initial system conditions are $y(0) = y'(0) = 0$.

Solve the equation, by Laplace transform, and/or CF/PI method, and/or using technology (either directly or to support the Laplace transform approach).

Identify the transient and steady state terms in the solution and, giving reasons, state whether the system described by the equation is over-damped or under-damped. Describe the behaviour of the system after a long time.

TABLE 17.2 LAPLACE TRANSFORM TABLES

$x(t)$	$X(s)$
k	$\dfrac{k}{s}$
kt	$\dfrac{k}{s^2}$
$\dfrac{kt^n}{n!}$	$\dfrac{k}{s^{n+1}}$
$ke^{-\alpha t}$	$\dfrac{k}{s+\alpha}$
$\dfrac{k}{\omega}\sin(\omega t)$	$\dfrac{k}{s^2+\omega^2}$
$k\cos(\omega t)$	$\dfrac{ks}{s^2+\omega^2}$
$\dfrac{k}{\alpha}(1-e^{-\alpha t})$	$\dfrac{k}{s(s+\alpha)}$
$kte^{-\alpha t}$	$\dfrac{k}{(s+\alpha)^2}$
$\dfrac{k}{2\omega^3}(\sin(\omega t)-t\cos(\omega t))$	$\dfrac{k}{(s^2+\omega^2)^2}$
$\dfrac{k}{(\omega^2-\Omega^2)}(\dfrac{1}{\Omega}\sin(\Omega t)-\dfrac{1}{\omega}\sin(\omega t))$	$\dfrac{k}{(s^2+\Omega^2)(s^2+\omega^2)}\quad \omega^2\neq\Omega^2$
$\dfrac{k}{(\beta-\alpha)}(e^{-\alpha t}-e^{-\beta t})$	$\dfrac{k}{(s+\alpha)(s+\beta)}\quad \alpha\neq\beta$
$\dfrac{k}{(\beta-\alpha)}(\beta e^{-\beta t}-\alpha e^{-\alpha t})$	$\dfrac{ks}{(s+\alpha)(s+\beta)}\quad \alpha\neq\beta$
$k[\dfrac{1}{(\beta-\alpha)(\gamma-\alpha)}e^{-\alpha t}+\dfrac{1}{(\alpha-\beta)(\gamma-\beta)}e^{-\beta t}$ $+\dfrac{1}{(\alpha-\gamma)(\beta-\gamma)}e^{-\gamma t})$	$\dfrac{k}{(s+\alpha)(s+\beta)(s+\gamma)}$ $\alpha\neq\beta\neq\gamma$

$k[\dfrac{-\alpha}{(\beta-\alpha)(\gamma-\alpha)}e^{-\alpha t}+\dfrac{-\beta}{(\alpha-\beta)(\gamma-\beta)}e^{-\beta t}$ $+\dfrac{-\gamma}{(\alpha-\gamma)(\beta-\gamma)}e^{-\gamma t})$	$\dfrac{ks}{(s+\alpha)(s+\beta)(s+\gamma)}$ $\alpha\neq\beta\neq\gamma$
$k[\dfrac{\alpha^2}{(\beta-\alpha)(\gamma-\alpha)}e^{-\alpha t}+\dfrac{\beta^2}{(\alpha-\beta)(\gamma-\beta)}e^{-\beta t}$ $+\dfrac{\gamma^2}{(\alpha-\gamma)(\beta-\gamma)}e^{-\gamma t})$	$\dfrac{ks^2}{(s+\alpha)(s+\beta)(s+\gamma)}$ $\alpha\neq\beta\neq\gamma$
$\dfrac{k}{(\alpha^2+\omega^2)}(e^{-\alpha t}-\cos(\omega t)+\dfrac{\alpha}{\omega}\sin(\omega t))$	$\dfrac{k}{(s+\alpha)(s^2+\omega^2)}$
$\dfrac{k}{(\alpha^2+\omega^2)}(-\alpha e^{-\alpha t}+\alpha\cos(\omega t)+\omega\sin(\omega t))$	$\dfrac{ks}{(s+\alpha)(s^2+\omega^2)}$
$\dfrac{k}{\omega}e^{-\beta t}\sin(\omega t)$	$\dfrac{k}{(s+\beta)^2+\omega^2}$
$ke^{-\beta t}\cos(\omega t)$	$\dfrac{k(s+\beta)}{(s+\beta)^2+\omega^2}$
$k(e^{-\beta t}\cos(\omega t)-\dfrac{\beta}{\omega}\sin(\omega t))$	$\dfrac{ks}{(s+\beta)^2+\omega^2}$
$\dfrac{k}{(\omega^2+(\alpha-\beta)^2)}(e^{-\alpha t}-e^{-\beta t}\cos(\omega t)$ $+\dfrac{\alpha-\beta}{\omega}e^{-\beta t}\sin(\omega t))$	$\dfrac{k}{(s+\alpha)((s+\beta)^2+\omega^2)}$
$\dfrac{k}{\omega}e^{-\beta t}\sin(\omega t)-\dfrac{k\alpha}{(\omega^2+(\alpha-\beta)^2)}(e^{-\alpha t}$ $-e^{-\beta t}\cos(\omega t)+\dfrac{\alpha-\beta}{\omega}e^{-\beta t}\sin(\omega t))$	$\dfrac{ks}{(s+\alpha)((s+\beta)^2+\omega^2)}$

Formulae for derivatives of functions

$\dfrac{dx}{dt}$	$sX(s)-x(0)$
$\dfrac{d^2x}{dt^2}$	$s^2X(s)-sx(0)-\dfrac{dx}{dt}(0)=s^2X(s)-sx(0)-x'(0)$

18

Taylor Series

"Visual mathematics: no magic bullets, but such tantalizing potential!" David Olson

18.1 THE ESSENTIAL ROLE OF TAYLOR SERIES

Often the mathematics needed to describe an engineering system is non-linear and therefore complicated: that is, functions more complicated than straight lines are required. However, it is useful sometimes to be able to use an approximate model of the situation which is linear - that is, described by straight lines. You can then make good use of your knowledge of straight lines.

Example 18.1.1

To see how this can work, consider Figure 18.1(a) – (c). This shows plots of the function $y(x) = x\sin(20x)e^{-0.2x}$ for $x \geq 0$, successively zooming in on the point $x = 1$ on the curve.

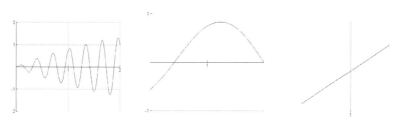

Figure 18.1(a)   Figure 18.1(b)   Figure 18.1(c)

You can see that as you zoom in more closely, the function appears progressively less curvy and, when magnified enough, the function eventually looks like a straight line. Engineers often use such thinking to obtain practical simplified linear laws which work locally – in time, space, or in whatever domain the model works.

If a straight line is not adequate, then we can extend this idea by using an appropriate higher order polynomial such as a quadratic or cubic, to give an approximate form of a function which works over the range we want. Polynomials are simple, manageable functions which can be built up in order to enable engineers to model complicated situations. One example of this approach, is the finite difference method as used in heat transfer, fluid flow and similar engineering areas.

These are the ideas which lie behind the Taylor series. You have the opportunity in this chapter to explore these important ideas a little further. To help you do that, **do the exercises and activities** as you come to them.

18.2 LINEARISATION

If we are to approximate a function locally by a straight line, say near to a point $x = a$, then it is necessary to work out how to find the equation of the right straight line in terms of the function we are approximating.

Take the function $y(x)$ close to $x = a$. We shall call the straight line which "fits" to the function $p_1(x)$. Thus:

$$y(x) \approx p_1(x) = c_0 + c_1 x \ . \tag{18.1}$$

The two questions we must ask are:

- For a given function at $x = a$, what are the right values of c_0 and c_1?

- How well does the line fit to the curve?

Take the first question first. To find c_0 and c_1, first make the line $p_1(x)$ fit exactly to $y(x)$ at the point $x = a$. Algebraically, this gives

$$p_1(a) = y(a) \approx c_0 + c_1 a \ . \tag{18.2}$$

Next, make the slope of the line $p_1'(x)$ fit exactly to the slope of the curve $y'(x)$ at the point $x = a$. Using the derivatives of $p_1(x)$ and $y(x)$, this gives:

$$p_1'(x) = y'(a) = c_1 \ . \tag{18.3}$$

You can then solve (18.2) and (18.3) to give the values of c_0 and c_1 in terms of $y(x)$ and $y'(x)$

$$\begin{aligned} c_0 &= y(a) - a y'(a) \\ c_1 &= y'(a) \end{aligned} \tag{18.4}$$

Thus if you know $y(x)$ you can find the straight line which "fits" it near to $x = a$.

Algebraically $y(x) \approx c_0 + c_1 x = y(a) - a y'(a) + y'(a)x$

or near $x = a$, $y(x) \approx y(a) + y'(a)(x - a) \ . \tag{18.5}$

This is called a *first order or linear approximation to y(x) about x = a*, or a *first order Taylor polynomial*.

The equation on the RHS represents a straight line, which has the same value and slope as $y(x)$ at $x = a$.

Example 18.2.1

Find a first order approximation to the function $y = e^x$ about $a = 0$.

Use the formula (18.5) with $a = 0$:

$$y(x) \approx y(0) + y'(0)(x - 0) \ . \qquad\qquad (18.6)$$

In this case $y(0) = e^0 = 1$ $\qquad\qquad\qquad\qquad$ (18.7)

and $\qquad\qquad y'(0) = e^0 = 1$ because $\dfrac{d}{dx}(e^x) = e^x$. $\qquad$ (18.8)

Putting (18.7) and (18.8) together into (18.6) then:

$$y(x) \approx 1 + 1(x - 0) = 1 + x \ .$$

It is useful that you can check visually that you have the right line, by plotting $y = e^x$ and $y = 1 + x$ on the same graph, as in Figure 18.2.

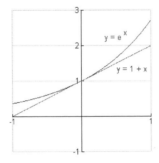

Figure 18.2 A straight line approximation to an exponential curve

To answer the second question, you can see that the fit is good "close to" $x = 0$ in both directions, as it was designed to be, but as x increases, the two curves move apart. Thus the linear approximation is good over a restricted range only.

In this particular case, where $a = 0$, the formula (18.5) becomes

$$y(x) \approx y(0) + y'(0)x \qquad\qquad\qquad (18.9)$$

and this is called a *first order Maclaurin polynomial*.

This is just a Taylor polynomial which will be a good fit near to the point x = 0. You might note that the Taylor polynomial is just a translation of the Maclaurin polynomial by a value of a in the x direction.

Example 18.2.2

A more complex function arises from Example 18.1.1. The principle is the same: the algebra is a little more complicated. We aim to find a linear approximation to $y(x) = x\sin(20x)e^{-0.2x}$ which is valid close to $a = 1$.

Using (18.5) again, but with $a = 1$ this time: $\qquad\qquad y(x) \approx y(1) + y'(1)(x - 1)$.

DO THIS NOW! 18.1

Show, using technology or otherwise, that $y(1) = 0.747456$ and $y'(1) = 7.28015$.

The line is then $y(x) \approx 0.747456 + 7.28015(x-1)$. This is plotted in Figure 18.3, and once more it is apparent that the fit is good near to the specified point $x = 1$.

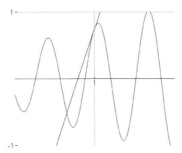

Figure 18.3 Taylor polynomial fit at $x = 1$.

DO THIS NOW! 18.2

Fit a **linear** approximation to each of the following as detailed. In each case plot the original function and the approximation and validate your calculations by observing how good the fit is:

(Note you must use RADIAN measure for the trigonometric functions)

1. Approximate $y(x) = \sin(x)$ close to $x = 0$.

2. Approximate $y(x) = \cos(x)$ close to $x = 0$.

3. Approximate $y(x) = e^x$ close to $x = 1$.

4. Approximate $y(x) = \sin(x)$ close to $x = \dfrac{\pi}{2}$.

5. Approximate $y(x) = e^{-2x}$ close to $x = 1$.

18.3 MACLAURIN SERIES
18.3.1 What is the Maclaurin series?

The idea of approximating one function locally by a simpler function can be extended and generalized from a linear approximation in order to represent the function by a series of powers of x: a so-called **power series or polynomial expansion about a point**. Thus near to $x = 0$ we could ask if a function $y(x)$ can be represented by a series of the form:

$$y(x) = c_0 + c_1 x + c_2 x^2 + c_3 x^3 + ... \qquad (18.10)$$

It turns out that this can be done for many common functions, and such a representation is called a **Maclaurin series**.

There are various points to make about the Maclaurin series, before going on to look at an example which will illustrate some of the points.

- The series is infinite and, in the cases where it works, it gets closer and closer to the function as more terms are added – in that case we say the series converges. Proving which series do converge is beyond this book – one needs to take care with infinite processes - but you do need to be aware that in some cases this approach does not work. Usually tables of series for standard functions will state when and where they work.

- As you add more terms, the series fits better and better to the function over a wider range, but the fit is best at and "near" $x = 0$. Taking only the first two terms gives the linear approximation derived in Section 18.2.

18.3.2 Finding the coefficients in the Maclaurin series

The coefficients $c_0, c_1, c_2, c_3, ...$ of the series depend on the function being approximated, and are given by:

$$c_0 = y(0)$$

$$c_1 = y'(0) = \frac{dy}{dx}(0)$$

$$c_2 = \frac{y''(0)}{2!} = \frac{1}{2!}\frac{d^2 y}{dx^2}(0)$$

$$c_3 = \frac{y'''(0)}{3!} = \frac{1}{3!}\frac{d^3 y}{dx^3}(0)$$

and in general
$$c_n = \frac{y^{(n)}(0)}{n!} = \frac{1}{n!}\frac{d^n y}{dx^n}(0) . \qquad (18.11)$$

Note that the general formula (18.11) works even for c_0 as long as you recall that $0!=1$. How do we justify these formulae? Here is the derivation of the first three, just to show you how it works.

To find c_0 start with (18.10) and set $x = 0$:

$$y(0) = c_0 + c_1 0 + c_2 0^2 + c_3 0^3 + ... = c_0 .$$

To find c_1 start with (18.10), differentiate once, and then set $x = 0$:

$$y'(x) = c_1 + 2c_2 x + 3c_3 x^2 + 4c_4 x^3 + ...$$

and so
$$y'(0) = c_1 + 2c_2 0 + 3c_3 0^2 + 4c_4 0^3 + ... = c_1 .$$

To find c_2 start with (18.10), differentiate twice, and then set $x = 0$:

$$y''(x) = 2c_2 + 3 \times 2c_3 x + 4 \times 3c_4 x^2 + ...$$

so
$$y''(0) = 2c_2 + 3 \times 2c_3 0 + 4 \times 3c_4 0^2 + ... = 2c_2$$

and hence $c_2 = \dfrac{y''(0)}{2!}$.

The pattern continues – to find c_n start with (18.10), differentiate n times, and then set $x = 0$.

You can tell from the formulae for the coefficients that this process is certainly only possible when the function can be differentiated as many times as necessary at some point on the function.

Example 18.3.1

Find the Maclaurin series for $y(x) = e^x$.

We shall find the first five terms using (18.11) and then look to see if a pattern is emerging. Note that the derivatives in this case are easy to find. Here they are up to the fourth derivative:

$$y(x) = y'(x) = y''(x) = y'''(x) = y^{(iv)}(x) = e^x$$

and so $y(0) = y'(0) = y''(0) = y'''(0) = y^{(iv)}(0) = e^0 = 1$.

Thus the coefficients are

$$c_0 = 1$$
$$c_1 = 1$$
$$c_2 = \frac{1}{2!}$$
$$c_3 = \frac{1}{3!}$$
$$c_4 = \frac{1}{4!} .$$

A pattern is emerging, that $c_n = \dfrac{1}{n!}$.

The series is then $e^x = 1 + x + \dfrac{x^2}{2!} + \dfrac{x^3}{3!} + \dfrac{x^4}{4!} + \ldots$.

To see how the series represents the function, you can plot $y(x) = e^x$, and also each successive approximation to it using the series, forming the so-called partial sums, adding up the first term, then the first two, then the first three, and so on. As more terms are added in, you can see how the shape of the series graph stays close to that of the function over a wider range. You can see this effect in Figure 18.4.

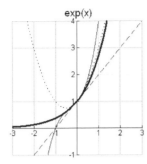

exp(x)

$$1 + x \qquad \text{-------}$$
$$1 + x + x^2 \qquad \text{...........}$$
$$1 + x + x^2 + x^3 \text{\underline{\hspace{2cm}}}$$

Figure 18.4 Maclaurin polynomials fit progressively better to the exponential as their order increases

DO THIS NOW! 18.3

Derive the Maclaurin series for the functions below (remember RADIANS!)

1. $y = \sin(x)$

2. $y = \cos(x)$

3. $y = e^{-2x}$

Convince yourself that they are correct by plotting the original functions and the partial sums of the series, and note how accurate the approximations are and for what ranges.

18.4 GETTING AWAY FROM $x = 0$: TAYLOR SERIES

The Maclaurin series gives successive approximations to $y(x)$ which work for values of x close to 0. It is useful to be able to generalise this idea by saying we want a series giving successive approximations to y which works for values of x close to any given value a. This more general form is called the *Taylor series*. It is formed by taking the result for the Maclaurin series and shifting it along the x-axis to $x = a$.

Combining (18.10) and (18.11), the Maclaurin series can be written:

$$y(x) = y(0) + y'(0)x + \frac{y''(0)}{2!} x^2 + \frac{y'''(0)}{3!} x^3 + ... \ . \qquad (18.12)$$

Moving this to $x = a$ leads to the Taylor series:

$$y(x) = y(a) + y'(a)(x-a) + \frac{y''(a)}{2!} (x-a)^2 + \frac{y'''(a)}{3!} (x-a)^3 + ... \ . \qquad (18.13)$$

If we let $h = (x - a)$, this is the distance of the point x away from the point a. It can then sometimes be useful to represent the series in terms of h, rewriting (18.13) in the form:

$$y(a+h) = y(a) + hy'(a) + \frac{h^2}{2!} y''(a) + \frac{h^3}{3!} y'''(a) + ... \qquad (18.14)$$

The next example shows a Taylor series in action.

Example 18.4.1

Find the Taylor series for $y = e^x$ near to $x = 1$.

From (18.13) the Taylor series is:

$$y(x) = y(a) + y'(a)(x-a) + \frac{y''(a)}{2!}(x-a)^2 + \frac{y'''(a)}{3!}(x-a)^3 + \ldots .$$

We want an expansion which works near to x = 1, so we take $a = 1$:

$$y(x) = y(1) + y'(1)(x-1) + \frac{y''(1)}{2!}(x-1)^2 + \frac{y'''(1)}{3!}(x-1)^3 + \ldots . \qquad (18.15)$$

To find the terms in the series, as with Example 18.3.1, we need

$$y(x) = y'(x) = y''(x) = y'''(x) = y^{(iv)}(x) = e^x ,$$

but this time we must work values out at $x = 1$:

$$y(1) = y'(1) = y''(1) = y'''(1) = y^{(iv)}(1) = e^1 = e .$$

Putting these values into (18.15) the Taylor series expansion around x = 1 is:

$$y(x) = e^x = e + e(x-1) + \frac{e}{2!}(x-1)^2 + \frac{e}{3!}(x-1)^3 + \ldots ,$$

which you can usefully rewrite as:

$$e^x = e(1 + (x-1) + \frac{(x-1)^2}{2!} + \frac{(x-1)^3}{3!} + \ldots) . \qquad (18.16)$$

Once more you can check the validity of your answer by plotting the original function and the partial sums of the series, gradually adding more terms, and see how good the fit is around $x = 1$. Figure 18.5 shows how good the approximation is close to $x = 1$, even if only the first three terms of the series are added in - a quadratic approximation to an exponential function!

$$e^x \approx 2.71828 + 2.71828(x-1) + 1.35914(x-1)^2$$

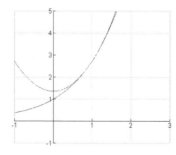

Figure 18.5 A quadratic fit to the exponential function near to $x = 1$

DO THIS NOW! 18.4

Find Taylor series for the functions as specified below, and explore the approximation graphically in each case.

1. $y = \sin(x)$ about $x = \dfrac{\pi}{2}$.

2. $y = \cos(2x)$ about $x = \pi$.

3. $y = e^{-2x}$ about $x = 2$.

18.5 USES OF TAYLOR AND MACLAURIN SERIES

Taylor and Maclaurin series have many important practical uses, not least because the ideas form an important part of evaluating solutions to field equations, for instance, for electric, magnetic or flow fields. There are many other mathematical uses, and we describe some of these below.

18.5.1 Approximations to functions

You have seen illustrations of this aspect in Sections 18.3 and 18.4. Do not forget that when you are dealing with infinite series, you must be sure that the series converge, and to find the series coefficients you need to be able to differentiate the function as many times as necessary and to evaluate it at a particular point.

You have seen that many common functions can be usefully expressed as polynomials for some range. In fact all computing machines and calculators use forms of these power series to work out values for common functions such as $y = \sin(x)$, $y = \cos(x)$, $y = e^x$ and so on.

18.5.2 Approximate values

A related application of power series is to find particular values of functions. For instance if you worked on the railways years ago, before the advent of calculators, you would need tables of values for $y = \sin(x)$; but for many measurements for the construction of railway tracks the angle x is small. This means that, as long as you work in radians, the Maclaurin series for $y = \sin(x)$ gives the approximation $\sin(x) \approx x$ (ignoring terms of x^3 and higher) and the tables are not needed.

DO THIS NOW! 18.5

Investigate the approximation $\sin(x) \approx x$ and find out how accurate it is. It is perfect at $x = 0$, but how big does x have to be before the error reaches 10%?

18.5.3 Indeterminate forms

An indeterminate form is one where substituting values in directly leaves uncertainty as to what the value is. One example is working out say $\dfrac{\sin(x)}{x}$ when $x = 0$. Direct substitution gives $\dfrac{0}{0}$, which as it stands is *indeterminate*. To investigate what happens when x gets close to 0, we can apply the Maclaurin series

$$\sin(x) = x - \frac{x^3}{3!} + \frac{x^5}{5!} - .... \text{ thus:}$$

$$\frac{\sin(x)}{x} = \frac{x - \frac{x^3}{3!} + \frac{x^5}{5!} - ...}{x} = 1 - \frac{x^2}{3!} + \frac{x^4}{5!} - ... .$$

You can then see that what happens as x approaches 0 is that $\frac{\sin(x)}{x}$ approaches

a value of 1 – the "divide by zero " problem has disappeared as x has been cancelled
out.

DO THIS NOW! 18.6

Use the appropriate Maclaurin series to establish the behaviour of some of the
following indeterminate forms in Figure 18.6 – noting as you go that some CAS
technology has this process built in.

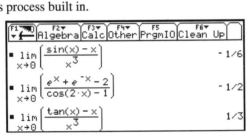

Figure 18.6

18.5.4 Establishing relationships between functions

There are many relationships between functions which can be established using
Maclaurin series.

Example 18.5.1

Use Maclaurin series to demonstrate that $\frac{d}{dx}(\sin(x)) = \cos(x)$.

Start with $\sin(x) = x - \frac{x^3}{3!} + \frac{x^5}{5!} - ....$ and differentiate it term by term - assuming it
is OK to do this with an infinite series!

$$\frac{d}{dx}(\sin(x)) = \frac{d}{dx}(x - \frac{x^3}{3!} + \frac{x^5}{5!} - \frac{x^7}{7!} + ...) = 1 - \frac{3x^2}{3!} + \frac{5x^4}{5!} - \frac{7x^6}{7!} + ...$$

$$= 1 - \frac{x^2}{2!} + \frac{x^4}{4!} - \frac{x^6}{7!} + ... = \cos(x)$$

as the last series is just the Maclaurin series for $y(x) = \cos(x)$.

DO THIS NOW! 18.7

Use similar methods to Example 18.5.1 to show that:

1. $\dfrac{d}{dx}(\cos(x)) = -\sin(x)$

2. $\sin(2x) = 2\sin(x)\cos(x)$

3. $\cos(2x) = \cos^2(x) - \sin^2(x) = 1 - 2\sin^2(x) = 2\cos^2(x) - 1$

DO THIS NOW! 18.8

One of the major mathematical relationships, introduced in Chapter 6, is Euler's relationship:

$$e^{jx} = \cos(x) + j\sin(x)$$

Establish this relationship by using the Maclaurin series

$$e^{jx} = 1 + jx + \frac{(jx)^2}{2!} + \frac{(jx)^3}{3!} + \dots ,$$

splitting it into real and imaginary parts, and relating it to the series for $y = \sin(x)$ and $y = \cos(x)$. (You will need to remember that $j^2 = -1$)

END OF CHAPTER 18 PROBLEMS – DO ALL THESE NOW!

1. In this question, work on paper and then confirm using CAS – you can see CAS output from one system in Figure 18.7.

   Determine the first 6 terms of the Maclaurin series for $y = \sin(x)$ about $x = 0$.

   Plot the graph of both the function and the series approximation.

   Hence determine the value of $\sin(0.3^c)$ correct to four decimal places.

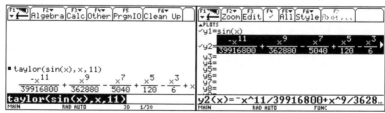

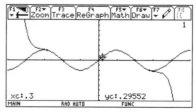

Figure 18.7 A Taylor series fit on CAS.

2. Determine the first 6 terms of the Maclaurin series for $y = \cos(x)$ expanded about $x = 0$.

   Plot a graph of both the function and the series approximation using the first 6 terms.

   Hence determine the value of $y = \cos(x)$ at $x = 0$. to four decimal places.

   Hence write down the series for $y = \cos(3x)$.

3. Determine the first five terms of the Maclaurin series for $y = \ln(1+x)$.

   Why do you think there is no Maclaurin series for the function $y = \ln(x)$? (Hint: plot this function and see how it behaves near to $x = 0$.)

   The infinite Maclaurin series for $y = \ln(1+x)$ is valid only for $|x| < 1$ - why do you think this is?

4. Complete the following table of standard Maclaurin series for the functions given, either by calculating them (you may already have done this for some) or by looking in any standard reference book on mathematical functions):

$y(x)$	Series
$\sin(x)$	$x - \dfrac{x^3}{3!} + \dfrac{x^5}{5!} - \dfrac{x^7}{7!} + ...$
$\cos(x)$	
$\tan(x)$	
e^x	
$\sin^{-1}(x)$	
$\ln(1+x)$	
$e^{\sin(x)}$	
$e^x \sin(x)$	

5. Determine the Taylor Series about a for $y = \cos(a+h)$ and hence determine $\cos(32°)$, correct to 6 decimal places.

   (Hint: start with equation (18.14), take $a = 30°$, and suitable value of h. Remember that although the angles here are specified in degrees you must do your workings in RADIANS otherwise formulae derived using calculus are not valid.)

6. Find the Taylor series expansion of $y = \tan(a+h)$ as far as the fifth power.

19

Statistics And Data Handling

"Oh, people can come up with statistics to prove anything." H Simpson

19.1 WHY ENGINEERS NEED DATA HANDLING SKILLS

Engineers need to understand the systems they must work with, and base their knowledge on the facts of the situation and not on prejudice or instinct. Statistics is about gathering, organising and using data to understand and manage those systems. There are two main aspects:

- *Descriptive statistics*: this is about organising, summarising and presenting data to understand a situation.

- *Inferential statistics*: this is about using data in rational ways to inform decision-making processes. It can involve using small samples of data to make inferences about larger systems about which you do not have full information.

Not every system is completely described and understood, so we must gather data (often partial) and use the information to try to understand what is needed. An engineer may need to play a role in managing a large project or company and to understand the economic side – wages, prices, availability of resources, etc. Alternatively it might be necessary for instance to ensure that the breaking strain of concrete beams is extremely unlikely to lie outside certain tolerance limits. In either case this raises the question of what data are to be gathered and what is to be done with the information which arises from the data.

Thus choosing the right pool of data from which to gather sample readings is a necessary skill. This pool is called the *population*. You must then decide which characteristics you are going to measure, and what the measure is to be – this measure is called a *random variable*. You must also decide what your strategy will be for gathering samples of data, taking due account of the fact that there will inevitably be variability between samples. This must be informed by what you intend to do with the data. It is not useful to gather data, and only *then* to start thinking about what to do with the information!

Having gathered the data, presentation matters. Presenting data in pictures is an excellent way to get to grips with the information, and choosing the best form of picture is an important skill. Sometimes, relationships between random variables can become clear only when the right picture is produced.

(A linguistic point – the word *"data"* is a plural word, coming from a Latin root. It is best equated with the word *"facts"*. Thus we would say *"data are"* etc. rather than *"data is"*. The singular of *"data"* is *"datum"*.)

This chapter will be restricted to discussing descriptive statistics and some of the basic techniques for handling data – including a simplified, initial approach to curve fitting. The full intricacies of inferential statistics and scientific decision-making are left for other books. The chapter is largely built around activities, mainly to be carried out on a spreadsheet. So **do the exercises and activities** as you go.

19.2 PRESENTING DATA IN PICTURES
19.2.1 Basic pictures

Presenting data in pictorial form can help you to understand what information the data hold, in a way that a page or file of numbers cannot. This section encourages you to explore the possible ways of representing the data graphically.

We assume that you are competent in using the basic graph plotting features of a spreadsheet such as Microsoft Excel. If you are not, then you must acquire this skill, as spreadsheets are now a standard part of the business and engineering toolkit.

You can choose from a variety of ways of presenting data pictorially, e.g. histogram, pie chart, bar chart, *x-y* graphs. This next exercise encourages you to explore the possibilities in your local spreadsheet.

DO THIS NOW! 19.1 Choosing a style of graph

The government regularly records many statistics, and one example is the average weekly hours worked by all persons. Recent data are shown in Table 19.1.

Year	2000	2001	2002	2003
Hours worked	23.8	24.2	25.0	25.9

Table 19.1 Average weekly hours worked by all persons March – May over a four year period

(Source: National Statistics Online at http://www.statistics.gov.uk/)

Type these data into your spreadsheet and explore how well the data are represented by different kinds of plot (pie chart, *x-y* graph, column chart, histogram).

What is the difference between the column chart and the histogram? Does one show you something the other does not?

What if you first make your "Hours worked" scale start from 22.5 (Excel defaults to this), and then force the scale to start at 0. Do the data tell you a different story then?

Authors' comments: a histogram has no gaps between "boxes" and it is the *area* of a box which reflects the number of values in that category, whereas with the column chart it is the *height* or length of the box. Of course these will have the same effect if all categories are of equal width, but not otherwise. The bar chart and histogram offer the same kind of information. See Figures 19.1 (a) and (b).

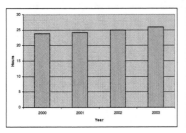
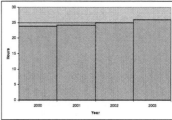

Figure 19.1 (a) Column chart   Figure 19.1 (b) Histogram

Excel note: produce the histogram in Figure 19.1 (b) by plotting the column graph, then right clicking the chart, and using Options on the Chart menu to change the gap between categories. Be precise with your terminology and adventurous with menus!

If you start your "vertical" scale at 22.5, it exaggerates the effect of any year on year changes; starting the scales at 0 gives a better indication of the scale of the changes.

The pie chart is not really helpful in this case.

19.2.2 Pictures comparing more than one data set

Pictures can play an important role in the initial exploration and comparison of data sets.

DO THIS NOW! 19.2 Comparing data sets

The data on average weekly working hours recorded by the government are also broken down between men and women, and recent data are presented in Table 19.2. It would be helpful to be able to plot these data in a way which allows comparison.

Year	2000	2001	2002	2003
Hours worked (men)	15.6	15.5	16.0	16.5
Hours worked (women)	8.2	8.7	8.9	9.3

Table 19.2 Average weekly hours worked by men and women March – May over a four year period

(Source: National Statistics Online at http://www.statistics.gov.uk/)

Type these data into your spreadsheet and explore how you might represent the three dimensional data to allow you to gain some information about the differences between hours worked by men and women. In Excel you might explore the use of an *x-y* graph or a 3-D column chart.

Does one representation show you something the other does not? Do you get useful information about the differences between hours worked by men and women, including the trends?

Authors' comments: Either picture gives some indication of the differences between hours worked by men and women, and the trends year on year.

19.2.3 The use of cumulative data

Sometimes the presentation of data category by category does not give a clear impression of what is going on, whereas if one were to look at the cumulative effect, say over a period of time, that may clarify what is happening.

DO THIS NOW! 19.3 Using cumulative graphs

Type the data in Table 19.3 below into your spreadsheet, and produce an *x-y* plot showing both sets of data on a single graph, perhaps by marking all three rows and choosing a *line graph*. Next arrange for the spreadsheet to reproduce the cumulative figures shown in Table 19.4, using formulae rather than calculating by hand. Now plot these cumulative figures on a similar single graph.

Which of the plots do you find most informative? Why?

Time period	1	2	3	4	5	6	7	8	9	10	11
Machine 1	39	86	97	102	74	53	62	69	80	53	48
Machine 2	45	67	61	77	71	51	45	58	48	62	36

Table 19.3 Production from two similar machines

Time period	1	2	3	4	5	6	7	8	9	10	11
Machine 1	39	125	222	324	398	451	513	582	662	715	763
Machine 2	45	112	173	250	321	372	417	475	523	585	621

Table 19.4 Cumulative production for data from Figure 19.2

19.3 SUMMARISING DATA SETS IN A FEW NUMBERS
19.3.1 Basic ideas

Original data often comes as a page (or pages) of numbers, which can be difficult to interpret. As you will have seen above, pictures can help in understanding the information contained in the data, but you can enhance your information by using what are called *summary statistics*: numbers which summarise certain aspects of the data. For simple data analysis these fall into two categories:

- *Measures of central tendency*, which give you some indication of where the "middle" of the data set is. These include the mean, the mode or modal group and the median.

- *Measures of spread or dispersion*, which give you some indication of how widely spread the data set is around its "middle". These include the variance and standard deviation, the range and the semi-interquartile range.

Each of these quantities is defined below, and they will all be illustrated in terms of the same data set, as presented in Example 19.1.

Example 19.1

The data in Table 19.5 concern items manufactured to a given nominal mass. A sample of 60 items has been chosen at random from the production line, and the mass of each, in kg, is recorded.

22.1	23.3	21.1	19.6	21.1	20.4
19.0	21.9	19.5	20.4	20.6	20.5
21.2	23.0	22.1	18.0	19.5	19.8
19.4	21.8	19.5	20.1	18.2	20.3
19.6	19.8	20.0	19.8	20.7	21.7
17.3	18.9	20.3	16.5	19.6	20.1
21.4	19.8	23.6	17.9	19.8	19.6
20.3	20.6	21.4	19.9	21.6	18.7
19.1	18.0	20.3	21.7	21.7	20.1
19.6	21.3	19.1	19.9	20.4	21.1

Table 19.5 Masses of a sample of 60 items from a production line, in kg,

Throughout this section the individual readings are labelled using subscripted variables $x_1, x_2, x_3, \ldots, x_N$, where N is the number of readings.

19.3.2 The mode (or modal group)

The *mode* tries to capture the idea of the most frequently occurring reading. With data such as that of Example 19.1, with its values given to 1 d.p., this idea is most meaningful if we group the data as in Table 19.6, and give the information as the *modal group* – the range of readings with the most values in it.

Thus we count up and record how many values fall into various ranges $16 - 16.9$, $17 - 17.9$ etc. The size of group is an issue here. A good rule of thumb is to keep things simple, and to have only about $6 - 9$ groups of equal size.

Class	16-16.9	17-17.9	18-18.9	19-19.9	20-20.9	21-21.9	22-22.9	23-23.9
No. of readings	1	2	5	19	15	13	2	3

Table 19.6

For Example 19.1 then, the modal group is $19 - 19.9$.

19.3.3 The median, range and semi-interquartile range

The *median* is defined as the "physical middle" of the data. That is, if all the readings are laid out in ascending or descending order, the median is the middle one. If there is an even number of readings it is the average of the two middle readings.

The data from Example 19.1, when sorted, look like Table 19.7 (we show the lowest, the highest, and the middle):

1	2	3	....	30	31	....	59	60
16.5	17.3	17.9	....	20.1	20.1	....	23.3	23.6

Table 19.7 Data from Example 19.1 sorted into ascending order.

As there is an even number of readings, the median is thus the average of readings 30 and 31:

$$\text{median} = \frac{20.1 + 20.1}{2} = 20.1 \;.$$

The *range* is one measure of spread, and is simply the difference between the highest and lowest values in the data set. For Example 19.1, the lowest value is 16.5 and the highest is 23.6, so the range is 7.1. This can be quite misleading about the scope of the data, as it can be seriously distorted by one extreme reading.

Another useful measure of spread is the **semi-interquartile** *range*. This uses the same idea as the median. Instead of working with the halfway point, it finds the two quarter-way points (or quartiles) and measures half the distance between those.

In Example 19.1, the lower quartile is reading number 15 in ascending order (19.5), and the upper quartile is reading number 45 in ascending order (21.1)

The semi-interquartile range is therefore $(21.1 - 19.5)/2 = 0.8$.

19.3.4 The mean, variance and standard deviation

A different approach to representing the centre and spread of a set of data uses the mean and standard deviation.

The **mean** or **average** is defined as the sum of all readings divided by the number of readings N:

$$\text{Mean} = \bar{x} = \frac{1}{N} \sum_{i=1}^{N} x_i \;. \tag{19.1}$$

In Example 19.1, $N = 60$, and the sum of all readings is 1213.6 (you can get your spreadsheet to do these calculations for you – and indeed probably to find the mean as well!). The mean is therefore $\bar{x} = \frac{1}{60}(1213.6) = 20.23$.

The first measure of spread or dispersion is the variance, and its square root, which is the standard deviation.

The idea of the **variance** is that it is the average value of the squared distance of each reading from the mean. Each distance is squared to remove the effect of any minus signs – we do not care whether a reading is above or below the mean, only how far from the mean it is. The **standard deviation** is used because by taking the square root of variance the measure is back in the same units as the original readings, whereas the variance is in (units)^2.

Sample variance

$$s^2 = \frac{1}{N} \sum_{i=1}^{N} [(x_i - \bar{x})^2] \;. \tag{19.2}$$

Sample standard deviation

$$s = \sqrt{\text{variance}} = \sqrt{\frac{1}{N} \sum_{i=1}^{N} [(x_i - \bar{x})^2]} \;. \tag{19.3}$$

There is a technical complication. To find an average you would normally divide by the number of readings N, and indeed if all you want is to find the variance for the sample set of readings you see then that is what you should do. However if you are using the sample variance and standard deviation to estimate the values for the *whole population* from which the sample is drawn, then you should divide not by N but by $(N - 1)$, as this will give a better estimate. Books on sampling theory give a full explanation of why this is. This measure is frequently distinguished from s by using s_{N-1}, and is the one which is most frequently used. For large samples (i.e. large values of N) it makes little difference whether you divide by N or by $(N - 1)$.

Thus *estimated population variance* is

$$s_{N-1}{}^2 = \frac{1}{N-1}\sum_{i=1}^{N}[(x_i - \bar{x})^2] \ . \tag{19.4}$$

Estimated population standard deviation is

$$s_{N-1} = \sqrt{\text{variance}} = \sqrt{\frac{1}{N-1}\sum_{i=1}^{N}[(x_i - \bar{x})^2]} \ . \tag{19.5}$$

In the case of Example 19.1, the variance is $s_{N-1}{}^2 = \frac{1}{59}(112.5273) = 1.91$.

The standard deviation is thus $s_{N-1} = 1.38$.

19.3.5 Comments

Note that the three measures of centre give different answers, but they each have their useful properties. For instance the median tends to play down the effect of extreme readings within data sets (a largest reading of 200 in Example 19.1 would have no more effect than the actual largest reading on the median. It would however have a distorting effect on the mean – where the actual value is taken into account rather than its position in ascending order).

Some machines work these *statistics* out as standard functions. For instance Excel has built-in functions AVERAGE, VAR and STDEV. Alternatively Figure 19.2 shows output from built-in statistics functions from a graphic calculator.

```
1-Var Stats          1-Var Stats
x=20.2250            ↑n=60.0000
Σx=1213.5000           minX=16.5000
Σx²=24654.9500         Q₁=19.5500
Sx=1.3773              Med=20.1000
σx=1.3657              Q₃=21.1500
↓n=60.0000             maxX=23.6000
```

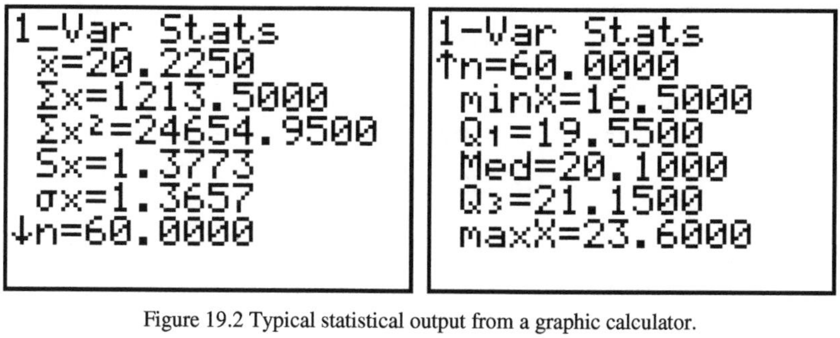

Figure 19.2 Typical statistical output from a graphic calculator.

DO THIS NOW! 19.4

Table 19.8 shows wage distribution data from a small engineering firm, in terms of the number of people on various hourly rates of pay.

Hourly rate (£)	4.50	8.00	10.00	15.00	60.00
Number of staff	6	21	19	3	1

Table 19.8 Data showing hourly pay rates and number of staff at those rates at a small engineering firm.

There are thus 50 members of staff, and $N = 50$. Note that the rate 4.50 occurs 6 times, and so on, and this must be taken into account in all calculations. Find:

(i) the mean hourly rate   (ii)   the modal group

(iii) the median hourly rate   (iv)   the standard deviation

(v) the range   (vi)   the semi-interquartile range.

You can check your answers in the box below.

Comment on why the mean and median are different, and which figure you would quote for comparison with other companies if you were a union negotiator!

Explain which version of standard deviation you have used and why. Comment on why the standard deviation and semi-interquartile range are different.

Author's notes on DO THIS NOW! 19.4 (and 19.3)

We get (i) £9.80; (ii) £8.00; (iii) £8.00; (iv) £7.53; (v) £55.50; (vi) £1.00.

We have used S_N because we have data for the whole population here.

The single person on £60 per hour distorts mean, standard deviation and range, but not median and semi-interquartile range (semi-IQR). This figure would not be directly relevant in comparisons with like workers, although it may have a bearing on the negotiating atmosphere!

There is a standard, and very useful graphical way of representing median, range and semi-IQR, using what is known as a box-and-whisker plot. This can also often make differences between two data sets apparent. For instance Figure 19.3 represents the data from DO THIS NOW 19.3! as two box-and-whisker plots, with plotted values from left to right representing lowest reading, lower quartile, median, upper quartile and highest reading.

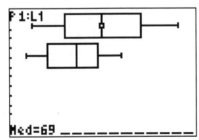

Figure 19.3 Box and whisker plots on a graphic calculator.

19.4 FITTING LAWS TO EXPERIMENTAL DATA

Sometimes with statistical analysis you are trying to find out if there is a relationship between two variables and, if so, to determine which mathematical law describes that relationship best. One good preliminary way to study what kind of relationship exists is to plot a graph of one variable against the other, and also to plot on the same graph any mathematical curves which may fit the data.

Initially one checks how well the mathematical law fits the data visually, but the scientific and objective way is to use the method of least squares. A full treatment of this lies beyond this book; here is a simple spreadsheet approach.

DO THIS NOW! 19.5

Table 19.9 shows the length of a steel wire under various loads.

Weight (N)	20	40	60	80	100
Length (m)	2.00042	2.00086	2.00132	2.00176	2.00219

Table 19.9

Place the data into the first two columns of your spreadsheet, and plot them as an *x-y* scatter graph (just points, not joined by lines, as they are individual readings).

You should be able to speculate that the points lie on a straight line and therefore the two variables are connected by a linear relationship, of the form $y = a + bx$

Estimate the values of *a* and *b* to give the "best" fit of the straight line to the data. First do this by eye, by making the third column of your spreadsheet hold the values of $a + bx$ for each value of *x*, and having *a* and *b* in cells on the spreadsheet. Add these values to your graph as a separate curve, joining points by lines to distinguish them from the measured data (your spreadsheet should look like Figure 19.4).

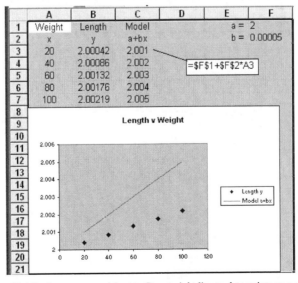

Figure 19.4 Setting up a spreadsheet to fit a straight line to data points on a graph.

Now play with the values of a and b on your spreadsheet to try to get the best fit you can between the data points and the line $y = a + bx$.

Let us remind you again however that this is not the scientific way of fitting a straight line, as two people might not agree on what is the best fit. The agreed method is called the *method of least squares*, which in many packages is implemented as a built-in function. For instance, in Excel one opens the graph by double clicking on it, selects the experimental data set by clicking once on any one data point, then from the Chart menu selects "Add trendline" and follows simple instructions, giving Figure 19.5. Other packages and technologies have similar built-in functions. Look at your technology to identify what it can do.

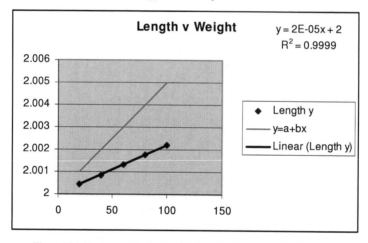

Figure 19.5 The result of using Excel's built-in "Insert trendline" feature.

In fact the "Add trendline" feature has a collection of curves which can be fitted to data according what "shape" the data has. See Figure 19.6.

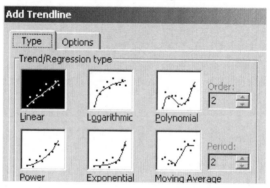

Figure 19.6 A range of curves can be fitted using "Insert trendline".

The trendline feature is using the method of least squares. To see how this works, you can carry on and have a go at DO THIS NOW! 19.6 You should find that it gives the same result as the trendline in DO THIS NOW! 19.5.

DO THIS NOW! 19.6

You can see how the method of least squares works by using the Solver feature of Excel as follows. Start by making your spreadsheet look like Figure 19.4. Make the links to the *a* and *b* cells live, so that as you change them they move the line. Get a good estimate of *a* and *b* by experiment.

Now, in the fourth column, calculate the values corresponding to

$$\text{(observed } y \text{ value} - \text{model } y \text{ value})^2$$

for each *x* value. These are the squared errors. Find the total of these values (the sum of squared errors). The idea is to minimize this sum of squared errors (hence the name least squares) by changing *a* and *b*. Experiment with changing *a* and *b* and you should see the sum changing – getting smaller as your line fits better.

Excel has a Solver function which will find the minimum of the sum of squared errors as *a* and *b* change. You may find this Solver on the Tools menu. The set-up is shown in Figure 19.7. Pressing Solve should make the line fit and give the same answer as the trendline. Make your spreadsheet do this now!

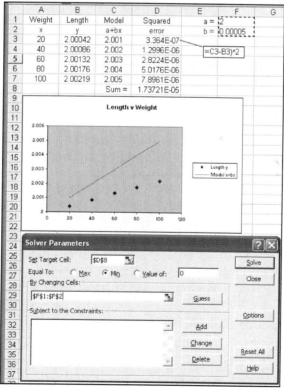

Figure 19.7 Setting up Excel's Solver to minimize squared errors.

We leave the topic of Statistics here. You should note that if you wish to take this topic further, you will probably need to learn other packages such as SPSS or SAS.

20

Probability

"Now what are the chances of that happening?" Harry Hill

20.1 WHAT IS PROBABILITY?
20.1.1 Probability, uncertainty and risk

We live in an uncertain world. The future can surprise us at any time. The fact that something has never happened does not mean that it will not do so in future. The fact that something has always been so does not mean it will be so forever. In the midst of this uncertainty, we have to live and make rational decisions. Probability theory is a mathematical attempt to quantify the chances of different events occurring. It is a crucial tool in engineering, whether attempting to quantify the risk of a piece of equipment failing so as to have a rational and safe replacement and renewal policy; assessing likelihood of producing and selling a poor quality item from a production line; or trying to design an investment policy which will manage movements in the stock market.

We all have a gut instinct about the chances of a particular event occurring, although some seem to get it right more often than others! In fact there are two ways this instinct comes about, and we must realise there are two kinds of probability:

a priori probability: this is where the probability can be reasoned out before the event (the meaning of the Latin phrase used). For example we can reason that a (fair!) coin is equally likely to come down heads or tails without tossing it.

a posteriori probability: this is where the probability can only be estimated in terms of what has gone before (*a posteriori* means "from behind"). A good example is the Fatal Accident Frequency Rate, in which the risk of future death in various activities (hang-gliding, ballroom dancing, working in the engineering industry) can be quantified in terms of historical accident rates, expressed as number of deaths per unit of time exposed to the risk. Of course this carries with it the possibility that if the situation has changed then rates from the past may no longer be a good predictor of the future.

Accepting all this, we state our feelings about probabilities in various ways, e.g.

"There's a 50-50 chance of getting a head when you toss a coin."

"I've got an even chance of getting a head when I toss a coin."

"The odds are 1 to 1 that I will get a head when I toss a coin."

Thus perhaps the first thing we need to do is agree to use a common language. Mathematicians define probability on a scale from 0 (no chance at all of something happening) to 1 (an event is certain to happen). Here is the definition:

20.1.2 Definition of Probability

If an event can occur in N distinct equally likely ways, and S of these ways possess some particular property A (we often think of this as corresponding to a "success"), then the probability of A occurring, denoted by $P(A)$ is defined as

$$P(A) = \frac{S}{N} = \frac{Number\,of\,successful\,outcomes}{Number\,of\,possible\,outcomes}\,. \qquad (20.1)$$

The number of "successful" outcomes cannot be less than 0, and since it also cannot be bigger than the total number of possible outcomes (!), then $0 \le S \le N$,

and so dividing through by N which is positive $0 \le \dfrac{S}{N} \le \dfrac{N}{N} = 1$.

That is $0 \le P(A) \le 1$ (20.2)

and probability lies on a scale from 0 to 1 as we said above.

You can see that if $S = 0$ then $P(A) = 0$ and this means no chance of success. Correspondingly, if $S = N$, then $P(A) = 1$ and all outcomes are successful.

You will find that using probability ideas in engineering situations does not make large demands on algebraic skills. Technology cannot help much, as the biggest difficulty is usually deciding what approach to take. That is, it is taking a problem which is often defined messily in words, extracting the relevant information and ignoring the irrelevant, and applying the right mathematical rules and ideas to get a sensible answer. In fact then, the best way to get to grips with the ideas of probability is to keep trying to apply them to different problems, so remember keep on **doing the exercises and activities**.

20.2 SIMPLE EXAMPLES – COMPLETE ENUMERATION

It is difficult to go straight into modelling probability in what can often be complicated engineering situations, so we begin with some simple and commonplace situations around which you can develop your thinking about probability. Simple situations can often be analysed by just writing down all the possible outcomes and calculating probabilities directly from there – so-called *complete enumeration*.

Example 20.2.1

Toss a coin once. There are two possible outcomes ($N = 2$), head (H) and tail (T). These are equally likely. Thus you can see simply that:

$$P(H) = \frac{S}{N} = \frac{1}{2} \qquad \text{and} \qquad P(T) = \frac{S}{N} = \frac{1}{2}.$$

This fits with the instinct that heads and tails are equally likely.

Example 20.2.2

Toss a coin twice. There are now four possible outcomes ($N = 4$), two heads (HH), two tails (TT), head followed by tail (HT) and tail followed by head (TH). Again these four are equally likely, and you can see simply that:

$$P(HT) = \frac{S}{N} = \frac{1}{4} \qquad \text{and} \qquad P(TT) = \frac{S}{N} = \frac{1}{4}, \text{ and so on.}$$

You can also see that $P(a \ head \ and \ a \ tail \ in \ any \ order) = \dfrac{S}{N} = \dfrac{2}{4} = \dfrac{1}{2}$

DO THIS NOW! 20.1

Use complete enumeration to analyse tossing a coin three times, calculating the probability of three heads HHH, of THT, and of two tails and a head in any order.

Example 20.2.3

Roll two fair six-sided dice and add the numbers showing. Analyse the probabilities of various outcomes by complete enumeration, for instance:

(i) event A = {scoring 12}   (ii) event B = {scoring 7}

(iii) event C = {scoring 3 or less}   (iv) event D = {scoring more than 3}.

All the possible outcomes (a complete enumeration of the sample space - there are $N = 36$ of them – count them!) are best presented as pairs of scores in a table:

1, 1	1, 2	1, 3	1, 4	1, 5	1, 6
2, 1	2, 2	2, 3	2, 4	2, 5	2, 6
3, 1	3, 2	3, 3	3, 4	3, 5	3, 6
4, 1	4, 2	4, 3	4, 4	4, 5	4, 6
5, 1	5, 2	5, 3	5, 4	5, 5	5, 6
6, 1	6, 2	6, 3	6, 4	6, 5	6, 6

Table 20.1

(i) $P(A) = P(\text{scoring } 12) = \dfrac{1}{36}$ (there is only one way of scoring 12).

(ii) $P(B) = P(\text{scoring } 7) = \dfrac{6}{36} = \dfrac{1}{6}$ (there are six ways of scoring 7 – see shaded cells in Table 20.1.)

(iii) $P(C) = P(\text{scoring 3 or less}) = \dfrac{3}{36} = \dfrac{1}{12}$ (there are three ways of scoring 3 or less – see bold cells in Table 20.1.)

(iv) $P(D) = P(\text{scoring more than 3})$.

You could count up the number of ways of scoring this as above, or you could realize that if there are three ways of scoring 3 or less, there must be $(36 - 3)$ ways of scoring more than 3. Thus

$$P(D) = \frac{33}{36} = \frac{11}{12}.$$

This example is picked up later in section 20.3.

DO THIS NOW! 20.2

When rolling two dice, calculate the probabilities of:

(v) making an odd numbered score;

(vi)   scoring 6 or 7.

20.3 MORE COMPLEX SITUATIONS – THE LAWS OF PROBABILITY

In many situations using complete enumeration can be complicated and long-winded, and in these cases it is useful to have laws for combining together probabilities of two or more events A, B, etc.

20.3.1 The Addition Law of probability

If A and B are two events then

$$P(A \text{ or } B) = P(A) + P(B) - P(A \text{ and } B). \qquad (20.3)$$

Note the use of the word "*or*" here. It is an example of how mathematicians can take a good English word and give it a slightly, but importantly different, meaning. It is the "*inclusive or*", in the sense that it means either A or B or *perhaps both* happen. This is as opposed to everyday usage which tends to be the "exclusive or": "I'm either having chips or pasta for lunch" (but presumably not both!)

One pictorial way to visualize this idea uses a Venn diagram as in Figure 20.2.

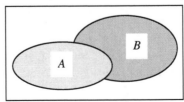

Figure 20.2 A Venn diagram showing events A and B which are not mutually exclusive

20.3.2 The Addition Law for **mutually exclusive events**

If the two events A and B cannot happen simultaneously, then we say they are *mutually exclusive*. In this case $P(A$ and $B) = 0$, and so the addition law becomes this:

If A and B are two mutually exclusive events then

$$P(A \text{ or } B) = P(A) + P(B) . \qquad (20.4)$$

The Venn diagram changes to represent this idea (Figure 20.3).

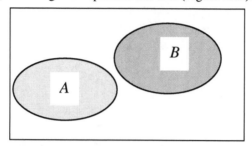

Figure 20.3 A Venn diagram showing events A and B which are mutually exclusive

Some examples will clarify the use of the law. We shall use the standard western deck of 52 playing cards to illustrate the law in action.

Example 20.3.1

Pick a single card from a deck of 52 ($N = 52$). First note some simple probabilities.

Event $K = \{$a King is drawn$\}$ $\qquad\qquad$ $P(K) = \dfrac{4}{52} = \dfrac{1}{13}$.

Event $Q = \{$a Queen is drawn$\}$ $\qquad\qquad$ $P(Q) = \dfrac{4}{52} = \dfrac{1}{13}$.

Event $H = \{$a heart is drawn$\}$ $\qquad\qquad$ $P(H) = \dfrac{13}{52} = \dfrac{1}{4}$.

Now using the addition law you can find the probabilities of various combinations of these events:

(a) Note that K and Q are mutually exclusive (a card cannot be both a King and a Queen!) and so

$$P(\text{King or Queen is drawn}) = P(K \text{ or } Q) = P(K) + P(Q) = \dfrac{1}{13} + \dfrac{1}{13} = \dfrac{2}{13} .$$

(b) Note that on the other hand K and H are *not* mutually exclusive – there is one card that can be both a King and a Heart (the King of Hearts!) - and so

$$P(\text{King or Heart is drawn}) = P(K \text{ or } H) = P(K) + P(H) - P(K \text{ and } H)$$

$$= \dfrac{1}{13} + \dfrac{1}{4} - \dfrac{1}{52} = \dfrac{16}{52} = \dfrac{4}{13} .$$

Note what is happening here. The figure of $\frac{1}{13}$ includes the King of Hearts, and so does the $\frac{1}{4}$, so to prevent this card being counted twice towards the probability, its probability $\frac{1}{52}$ must be subtracted once. The law thus makes sense in terms of the complete enumeration of the situation.

DO THIS NOW! 20.3

In the situation in which a single card is drawn, calculate the probabilities that:

(i) a red card is drawn.   (ii) a 2 or 3 is drawn   (iii) a 2 or a red card is drawn.

20.3.3 Complementary events

This is a natural point at which to pick up an issue which arose at the end of Example 20.2.3. First we need a bit of notation: If A is an event, then denote by $\overline{A}$ the event that A does not happen (many people read this as "not A"). A and $\overline{A}$ are called *complementary* events.

A Venn diagram gives a good way to represent this idea, as in Figure 20.4.

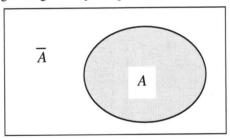

Figure 20.4 A Venn diagram showing complementary events.

Now clearly A and $\overline{A}$ are mutually exclusive (A cannot both happen and not happen simultaneously). Also, either A or $\overline{A}$ must happen (either A happens or it does not). These facts can be interpreted in mathematical statements:

$$P(A \text{ or } \overline{A}) = P(A) + P(\overline{A}) \qquad \text{(mutually exclusive).}$$

$$P(A \text{ or } \overline{A}) = 1 \qquad \text{(either } A \text{ or } \overline{A} \text{ must happen).}$$

Putting these together gives

$$P(A) + P(\overline{A}) = 1$$

or $\qquad P(A) = 1 - P(\overline{A})$.   (20.5)

It is often easier when finding the probability that A happens, to find first the probability of $\overline{A}$, and then use (20.5) to find $P(A)$, as illustrated in the next example.

Example 20.3.2

Return to Example 20.2.3, where two dice are rolled. Suppose you want to find the probability of scoring 11 or less. There are many combinations which produce a score of 11 or less, but working with the complementary event makes things simpler:

Let $\quad A = \{\text{scoring 11 or less}\}$

The event complementary to this is:

$$\overline{A} = \{not \text{ scoring 11 or less}\} = \{\text{scoring more than 11}\} = \{\text{scoring 12}\}$$

This is easier to work with as there is only 1 way out of 36 to score 12:

$$P(\overline{A}) = \frac{1}{36} \text{ and so } P(A) = 1 - P(\overline{A}) = 1 - \frac{1}{36} = \frac{35}{36}.$$

DO THIS NOW! 20.4

Use the complementary event approach to calculate the probability that a coin tossed four times does not show the same result on all four tosses. (We get 7/8)

20.3.4 The Multiplication Law of probability

The multiplication law deals with the probability of two events both happening. First you need the idea of a *conditional* probability. If A and B are two events then denote by $P(B/A)$ the probability that B happens given that A happens.

Then the multiplication law is this:

If the probability of A happening is $P(A)$, and the probability of B then happening given that A happens is $P(B/A)$, then the probability that both A and B happen is

$$P(A \text{ and } B) = P(A).P(B/A). \qquad (20.6)$$

Example 20.3.3

Two cards are drawn from a pack of 52 without replacement. What is the probability that both are clubs?

Let $A = \{\text{the first card is a club}\}$, and let $B = \{\text{the second card is a club}\}$.

Then $\quad P(A) = \dfrac{13}{52} = \dfrac{1}{4}$ $\qquad$ (there are 13 clubs out of 52)

and $\quad P(B/A) = \dfrac{12}{51} = \dfrac{4}{17}$ $\qquad$ (there are 12 clubs out of 51 cards left).

It is clear here that A affects B since when the second card is picked, the first card (which was a club) has already gone and has not been replaced. Finally then:

$$P(\text{both cards are clubs}) = P(A \text{ and } B) = P(A).P(B/A) = \frac{1}{4} \times \frac{4}{17} = \frac{1}{17}.$$

You could also find this by complete enumeration, writing down all the ways in which the two drawn cards can come out, but using the law is far simpler.

20.3.5 The Multiplication Law for independent events

In some cases the second event is not influenced by the first. The events are then called *independent*. In this case $P(B/A) = P(B)$, and the multiplication law becomes simpler.

If A and B are independent events then

$$P(A \text{ and } B) = P(A).P(B) . \qquad (20.7)$$

Example 20.3.4

Return to Example 20.3.2 where a coin is tossed twice.

Let $A = \{$first toss yields a head$\}$ and let $B = \{$second toss yields a head$\}$.

One would assume that each toss of the coin is independent of any previous tosses, so A and B are independent events. Thus

$$P(\text{getting two heads in two tosses}) = P(A \text{ and } B) = P(A).P(B) = \frac{1}{2} \times \frac{1}{2} = \frac{1}{4} .$$

Note that this agrees with the result which came from complete enumeration back in Example 20.2.2 – which is reassuring!

DO THIS NOW! 20.5

Two cards are drawn from a pack without replacement. What is the probability that both are kings? (We get 1/221. What do you get?)

Answer the same question in the situations where the number of cards drawn is three, then four, then five. (We get 1/5525, 1/270725, 0. What do you get?)

20.4 TREE DIAGRAMS

It has been a recurring theme of this book that a picture paints a thousand words. In dealing with probability, the Venn diagram can help with getting a picture of some of the concepts, but one of the most useful tools in helping to apply the laws correctly is the *tree diagram*, a graphical technique which helps you to sort out the realities of a situation and apply the correct law at the correct time. It is best introduced by an example.

Example 20.4.1

A box contains 14 light bulbs. 12 of these work and 2 do not. Someone comes in and picks two out at random to use.

What is the probability that (a) neither works, and (b) just one works?

Let $A = \{$the first one works$\}$ and let $B = \{$the second one works$\}$. This first step of setting up the problem mathematically by making definitions of the events can be under-emphasized, and can be a source of some difficulty.

For instance if you try to build an image of the situation by thinking in terms of taking out both bulbs simultaneously then that makes it more difficult. It is easier to see how taking two bulbs affects the probabilities, and entirely equivalent, to think of it as taking out the bulbs one after the other.

You can then analyse this situation using a tree diagram as shown in Figure 20.5. The tree diagram is a graphical way of forcing you to think the events through step by step. See if you can interpret the diagram, with the help of the comments below.

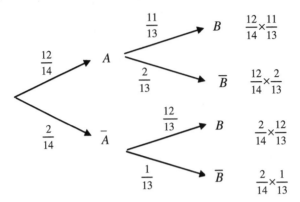

Figure 20.5 A tree diagram for Example 20.4.1.

The "tree" grows from left to right. You can see that all probabilities are thought out step by step, so any issues of independence must be dealt with as you go. Also each path followed along "branches" from left to right is separate, so mutual exclusivity is automatically dealt with.

As you travel along a path from left to right, you take in one probability **AND** then the next, so you **multiply** probabilities as you go, and get the overall probability of arriving at the end of that path. Then you choose the path(s) that you are interested in, and since these will be one **OR** the other, you *add* the resultant probabilities from each path you require. A good check is that all the probabilities down the right hand side must add up to 1, as you have to end up somewhere on the right hand side.

The answers to our particular questions are then:

(a) Only the bottom branch relates to the event that neither works, so:

$$P(\text{neither works}) = \frac{2}{14} \times \frac{1}{13} = \frac{1}{91}.$$

(b) Both middle branches have just one bulb working, so include both and add them.

$$P(\text{just one works}) = \frac{12}{14} \times \frac{2}{13} + \frac{2}{14} \times \frac{12}{13} = \frac{12}{91} + \frac{12}{91} = \frac{24}{91}.$$

DO THIS NOW! 20.6

Extend the tree diagram of Example 20.4.1 to deal with the situation in which three bulbs are taken. Answer the equivalent questions. That is, calculate the probabilities that (a) none of the three bulbs works and (b) just one of the three works.

Hints: Your tree will need three left-to-right layers so leave enough space for that! Some branches may come to an end before the third layer. Why?

We get (a) 0 and (b) 33/91. What do you get?

20.5 SOME MORE PROBABILITY PROBLEMS

Finally, here are some probability problems which bring in more realistic and engineering-related situations.

Example 20.5.1

Valves control gas flow through a pumping station. They are combined in various arrangements to increase safety by building in back-ups. These valves have a certain probability of staying open when the signal is sent to close ("failing open") and of failing to close when the signal is sent to open ("failing closed"). Here is an analysis of the probabilities that some configurations of valve will fail either to stop gas or allow it through when ordered to do so.

Suppose for each individual valve:

P(failing open) = 0.02   and P(failing closed) = 0.03

Example 20.5.1 (a) Two valves in series (Figure 20.6)

Figure 20.6 Gas valves in series.

Let A = {gas stops flowing when both valves are open and are ordered to close}

To find $P(A)$ directly is complicated as there are several ways in which the gas could stop, with either of the individual valves, or perhaps both valves working. However there is only one way that the gas could continue to flow, so this is a good chance to use the complementary event

$\overline{A}$ = {gas still flows when both valves are open and are ordered to close}.

$P(\overline{A})$ = P(both valves fail open) = P(first fails open *and* second fails open)

= P(first fails open).P(second fails open) = 0.02×0.02 = 0.0004.

Finally $P(A) = 1 - P(\overline{A}) = 1 - 0.0004 = 0.9996.$

Example 20.5.1 (b) Two valves in parallel (Figure 20.7)

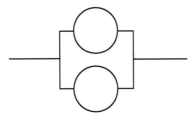

Figure 20.7 Gas valves in parallel.

Let B = {gas stops flowing when both valves are open and are ordered to close}

This time the head-on approach is the natural one, as there is only one way in which the gas will stop – that is, if both valves work.

Thus $P(B) = P(\text{both valves work}) = P(\text{first valve works } and \text{ second valve works})$

$$= P(\text{first valve works}) \times P(\text{second valve works}).$$

You can get these probabilities that each valve works by simply subtracting from 1 the probability that the valve fails open.

Thus $P(B) = (1 - 0.02) \times (1 - 0.02) = 0.98 \times 0.98 = 0.9604.$

DO THIS NOW! 20.7

Apply the ideas of Example 20.5.1 to the configuration in Figure 20.8.

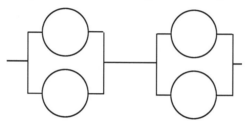

Figure 20.8 A more complicated (and realistic) valve arrangement.

(a) Find the probability that gas continues to flow when all valves are open and are ordered to close.

(b) Find the probability that gas flows when each valve is closed but is ordered to open.

(Hint: break the problem down - that is, find the probability for each parallel subgroup, then combine them together in series.)

We get (a) 0.001568, and (b) 0.998201. What do you get?

Example 20.5.2

In a factory, a piece of equipment will function only if 3 components A, B, and C are all working. The probability of A failing during one year is 0.05, that of B failing is 0.01, and that of C failing is 0.10. What is the probability that the equipment fails before the end of the year?

DO THIS NOW! 20.8

Try to finish Example 20.5.2!

Try first without reading these following hints, but if you are having problems, read the next paragraph and then have another go.

Hint: Once more, if you try to tackle this problem head-on, it will be complicated, as there are many ways in which the equipment can fail, but only one in which it can function properly. So deal with this problem by finding the probability that the equipment does not fail by the end of the year, and then use the complementary event rule to answer the original question.

We get 0.15355. Do you agree?

DO THIS NOW! 20.9

Try to tackle this problem using the methods you have been learning about. Some hints appear at the end of this box if you are having trouble getting started.

Metal rods are manufactured at a factory to a given specification - that they must be of length 40 ± 0.2 cm. Rods are made on one of three nominally identical machines A, B, and C in the factory. Machine A makes 43% of the rods, machine B makes 35%, and machine C makes the remainder. Of the rods made on machine A, 5% lie outside the tolerance limits; for machine B this figure is 4%; and for machine C it is 3%.

(i) Calculate the probability that a rod picked at random from the factory's output lies outside the tolerance limits.

(ii) Given that a rod lies outside the tolerance limits, calculate the probability that it was made on machine C.

Hints:

For part (i), use a tree diagram, with two layers of branches, one split by A, B, C, and one for good/bad in each of those cases.

For part (ii), use the multiplication law $P(A \text{ and } B) = P(A).P(B/A)$, but in rearranged

form $$P(B / A) = \frac{P(A \text{ and } B)}{P(A)} \ .$$

Choose your A and B carefully in this law to fit the problem.

We get (i) 0.0421 and (ii) 0.15677. Do you agree?

20.6 WHERE NEXT WITH PROBABILITY?

When you construct a tree diagram, or do anything else to find all the relevant probabilities in a situation, you are constructing what is called a *probability distribution*. This effectively means coming up with a formula or table which gives all probabilities in a given situation.

Often, you will find that the same mathematical distributions of probability will occur in very different practical situations, so these are given names – for instance, binomial, Poisson, Normal, uniform and exponential distributions. Study of these would be the next logical step in your learning about probability, but they are beyond the scope of this book.

If you have managed to do all of the **DO THIS NOW! 20.1-9**, exercises you should be feeling a little more confident in your understanding of probability now. Go back and do them now if you haven't already done so!

Glossary

In the pages which follow we have gathered together a collection of main facts and helpful formulae and terms. We hope you find this useful.

G1 GREEK ALPHABET

A	α	alpha	N	ν	nu
B	β	beta	Ξ	ξ	xi
Γ	γ	gamma	O	o	omicron
Δ	δ	delta	Π	π	pi
E	ε	epsilon	P	ρ	rho
Z	ζ	zeta	Σ	σ	sigma
H	η	eta	T	τ	tau
Θ	θ	theta	Y	υ	upsilon
I	ι	iota	Φ	ϕ	phi
K	κ	kappa	X	χ	chi
Λ	λ	lamda	Ψ	ψ	psi
M	μ	mu	Ω	ω	omega

G2 SI UNITS

The International System of units (SI units) is an internationally agreed set of units and symbols for measurement.

The base units for this system are metre, second and kilogram (symbols m, s, kg respectively).

Prefixes may be added, for instance:

c centi (10^{-2}) m milli (10^{-3}) μ micro (10^{-6}) k kilo (10^{3}) M mega (10^{6}).

Many software packages and machines (see for instance screen dumps in Figure G1) have settings available to make sure you work in the appropriate units. Make sure you know what your technology does.

Figure G1 Graphic calculator screens for specifying units.

G3 COMMON GRAPHS TO NOTE
G3.1 Identify which graph goes with which equation.

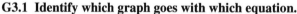

$$y = mx + c$$
$$y = x^2$$
$$y = \frac{1}{x}$$
$$y = |x|$$

G3.2 Identify which graph goes with which equation.

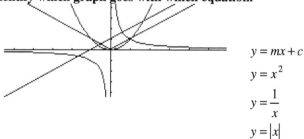

$$y = e^x$$
$$y = e^{-x}$$
$$y = \ln(x)$$

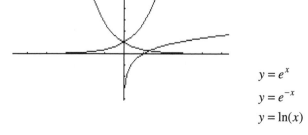

G3.3 Identify which graph goes with which equation.

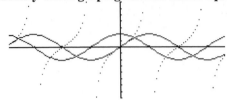

$$y = \sin(x)$$
$$y = \cos(x)$$
$$y = \tan(x)$$

G3.4 Identify which graph goes with which equation.

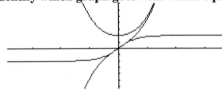

$$y = \sinh(x)$$
$$y = \cosh(x)$$
$$y = \tanh(x)$$

G3.5 Identify which graph goes with which equation.

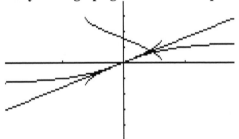

$$y = \sin^{-1}(x) = \arcsin(x)$$
$$y = \cos^{-1}(x) = \arccos(x)$$
$$y = \tan^{-1}(x) = \arctan(x)$$
$$y = x$$

G4 POWER SERIES
Maclaurin's Series:

$$f(x) = f(0) + xf'(0) + \frac{x^2}{2!}f''(0) + \frac{x^3}{3!}f'''(0) + \ldots$$

Taylor's Series:

$$f(a+h) = f(a) + hf'(a) + \frac{h^2}{2!}f''(a) + \cdots + \frac{h^{n-1}}{(n-1)!}f^{n-1}(a) + \frac{h^n}{n!}f^n(c)$$

where $a \le c \le a+h$.

Common Series

$$(1+x)^n = 1 + nx + \frac{n(n-1)}{2!}x^2 + \frac{n(n-1)(n-2)}{3!}x^3 + \dots$$

$$|x| < 1$$

$$(1+x)^{-1} = 1 - x + x^2 - x^3 + \dots$$

$$|x| < 1$$

$$e^x = 1 + x + \frac{x^2}{2!} + \frac{x^3}{3!} + \dots$$

All x

$$\ln(1+x) = x - \frac{x^2}{2!} + \frac{x^3}{3!} - \frac{x^4}{4!} + \dots$$

$$-1 \le x \le 1$$

$$\sin x = x - \frac{x^3}{3!} + \frac{x^5}{5!} - \frac{x^7}{7!} + \dots$$

All x

$$\cos x = 1 - \frac{x^2}{2!} + \frac{x^4}{4!} - \frac{x^6}{6!} + \dots$$

All x

$$\tan x = x + \frac{x^3}{3} + \frac{2x^5}{15} + \frac{17x^7}{315} + \dots$$

$$|x| < \frac{\pi}{2}$$

$$\arcsin x = x + \frac{1}{2}\frac{x^3}{3} + \frac{1 \times 3}{2 \times 4}\frac{x^5}{5} + \frac{1 \times 3 \times 5}{2 \times 4 \times 6}\frac{x^7}{7} + \dots$$

$$|x| < 1$$

$$\arccos x = \frac{\pi}{2} - \arcsin x$$

$$\arctan x = x - \frac{x^3}{3} + \frac{x^5}{5} - \frac{x^7}{7} + \dots$$

$$|x| < 1$$

$$\sinh x = x + \frac{x^3}{3!} + \frac{x^5}{5!} + \frac{x^7}{7!} + \dots$$

All x

$$\cosh x = 1 + \frac{x^2}{2!} + \frac{x^4}{4!} + \frac{x^6}{6!} + \dots$$

All x

$$\tanh x = x - \frac{x^3}{3} + \frac{2x^5}{15} - \frac{17x^7}{315} + \dots$$

$$|x| < \frac{\pi}{2}$$

$$\operatorname{arcsinh} x = x - \frac{1}{2}\frac{x^3}{3} + \frac{1 \cdot 3}{2 \cdot 4}\frac{x^5}{5} - \frac{1 \cdot 3 \cdot 5}{2 \cdot 4 \cdot 6}\frac{x^7}{7} + \dots$$

$$|x| < 1$$

$$\operatorname{arccosh} x = \ln 2x - \frac{1}{2}\frac{1}{2x^2} - \frac{1 \cdot 3}{2 \cdot 4}\frac{1}{4x^4} - \frac{1 \cdot 3 \cdot 5}{2 \cdot 4 \cdot 6}\frac{1}{6x^6} - \dots$$

$$x \ge 1$$

G5 COMMON NOTATION

$>$	greater than
$\geq$	greater than or equal to
$<$	less than
$\leq$	less than or equal to
a^n	a to the power n
$\sqrt{a}$	the non negative square root of a, where $a \geq 0$
$\sqrt[n]{a}$	the nth root of a, where $a \geq 0$
a^{-n}	the n th root of a where $a \neq 0$
$a^{m/n}$	$\sqrt[n]{a^m}$, where $n > 0$ and $a \geq 0$
∞	infinity
(x, y)	the Cartesian coordinates of a point
$\approx$	approximately equal to
$\sin^2(x)$	$(\sin(x))^2$ a peculiar mathematical notation
$[a,b]$	set of all numbers between a and b, including a and b
(a,b)	set of all numbers between a and b, not including a and b
$[a,b)$	set of all numbers between a and b, including a but not b
$|x|$	the absolute value of x
e	base for the natural logarithm function and the exponential function ($e \approx 2.718281$)
exp	the exponential function
$\arccos(x)$	angle in interval $[0, \pi]$ whose cosine is x. ($\cos^{-1}(x)$)
$\arcsin(x)$	angle in interval $[-\frac{\pi}{2}, \frac{\pi}{2}]$ whose sine is x. ($\sin^{-1}(x)$)
$\arctan(x)$	angle in interval $[-\frac{\pi}{2}, \frac{\pi}{2}]$ whose tangent is x. ($\tan^{-1}(x)$)
$\log_a$	logarithm function to base a
ln	natural logarithm function to $\log_e$ where $e \approx 2.718281$
j	$\sqrt{-1}$
Re(z)	the real part of z
Im(z)	the imaginary part of z
$\bar{z}$ or z*	the complex conjugate of z
$|z|$	the modulus of the complex number z
$\arg(z)$	the argument of the complex number z
(r, θ)	the polar form of a complex number, where r is the modulus and θ is an argument.

f' or $\dfrac{df}{dx}$ the (first) derivative of the function f with respect to x.

f'' or $\dfrac{d^2 f}{dx^2}$ the (second) derivative of the function f with respect to x.

$\displaystyle\int f(x)dx$ the indefinite integral of $f(x)$ with respect to x

$\displaystyle\int_a^b f(x)dx$ the definite integral of $f(x)$ with from a to b

$y(a) = b$ the condition that $y = b$ when $x = a$

A   a matrix

AB   the product of matrices **A** and **B**

A^n the n th power of the square matrix **A**

A + **B**   the sum of the matrices **A** and **B**

A − **B**   the difference of the matrices **A** and **B**

k**A**   the scalar multiple of the matrix **A** by a real number k

I   the identity matrix

A^{-1} the inverse of the invertible matrix **A**

det **A**   the determinant of the square matrix **A**

a_{ij} the element in the i th row and j th column of the matrix **A**

Ax = **b**   the matrix form of a set of simultaneous equations

0   the zero matrix.

G6 TABLE OF TRIGONOMETRIC FUNCTION FORMULAE

$$\sin^2\theta + \cos^2\theta = 1 \qquad\qquad \sec^2\theta = 1 + \tan^2\theta \qquad\qquad \mathrm{cosec}^2\theta = 1 + \cot^2\theta$$

$$\sin(A+B) = \sin A\cos B + \cos A\sin B \qquad\qquad \sin(A-B) = \sin A\cos B - \cos A\sin B$$

$$\cos(A+B) = \cos A\cos B - \sin A\sin B \qquad\qquad \cos(A-B) = \cos A\cos B + \sin A\sin B$$

$$\tan(A+B) = \frac{\tan A + \tan B}{1 - \tan A\tan B} \qquad\qquad \tan(A-B) = \frac{\tan A - \tan B}{1 + \tan A\tan B}$$

$$\sin 2A = 2\sin A\cos A \qquad\qquad \cos 2A = \cos^2 A - \sin^2 A = 1 - 2\sin^2 A = 2\cos^2 A - 1$$

$$\tan 2A = \frac{2\tan A}{1 - \tan^2 A}$$

Glossary

$$\sin A + \sin B = 2\sin\left(\frac{A+B}{2}\right)\cos\left(\frac{A-B}{2}\right)$$

$$\sin A - \sin B = 2\cos\left(\frac{A+B}{2}\right)\sin\left(\frac{A-B}{2}\right)$$

$$\cos A + \cos B = 2\cos\left(\frac{A+B}{2}\right)\cos\left(\frac{A-B}{2}\right)$$

$$\cos A - \cos B = -2\sin\left(\frac{A+B}{2}\right)\sin\left(\frac{A-B}{2}\right)$$

$$\sin A \cos B = \frac{1}{2}\{\sin(A+B) + \sin(A-B)\}$$

$$\sin A \sin B = \frac{1}{2}\{\cos(A-B) - \cos(A+B)\}$$

$$\cos A \cos B = \frac{1}{2}\{\cos(A+B) + \cos(A-B)\}$$